THE GEORGE FISHER BAKER
NON-RESIDENT LECTURESHIP
IN CHEMISTRY AT
CORNELL UNIVERSITY

NMR of Proteins
and Nucleic Acids

NMR of Proteins and Nucleic Acids

Kurt Wüthrich

Institute of Molecular Biology and Biophysics
Swiss Federal Institute of Technology
Zürich, Switzerland

A Wiley-Interscience Publication

John Wiley & Sons

New York • Chichester • Brisbane • Toronto • Singapore

Library of Congress Cataloging in Publication Data:

Wüthrich, Kurt.
    NMR of proteins and nucleic acids.

    (Baker lecture series)
    "A Wiley-Interscience publication."
    Bibliography: p.
    Includes index.
    1. Nuclear magnetic resonance spectroscopy.
2. Proteins—Analysis.   3. Nucleic acids—Analysis.
I. Title. II. Series. [DNLM: 1. Nuclear Magnetic
Resonance—methods.   2. Nucleic Acids—analysis.
3. Proteins—analysis.   QU 55 W973n]

QP519.9.N83W87   1986   574.19′245   86-7834
ISBN 0-471-82893-9

Printed in the United States of America

10 9 8 7 6

Preface

The structure and function of proteins and nucleic acids are a long established, central theme of biological sciences, which has gained added interest
by the fact that these biopolymers can now be modified efficiently to search
for improved or new biological activities. Considering that the number of
different variations for polynucleotide and polypeptide chains with n building blocks are, respectively, 4^n and 20^n, systematic screening of biopolymer
sequences must be discarded as a practicable strategy for using genetic
engineering or chemical synthesis in the design of new, improved compounds. More promising approaches are guided by basic knowledge of correlations of structure, conformation, and function of nucleic acids and proteins. This book treats the use of nuclear magnetic resonance for obtaining
such information.

The potentialities and practice of NMR studies with proteins and nucleic
acids were decisively changed during the past few years, so that perhaps the
time has come to give a comprehensive introduction to the underlying principles and experimental procedures. It is the purpose of this volume to furnish
such an account. In writing this book I have had in mind practicing scientists
and students of biochemistry, chemistry, biophysics, and molecular biology
who are concerned in their daily work with the structure and function of
proteins and nucleic acids. This group can be expected to represent the
primary users of the results of NMR studies on biopolymers. It is my hope
that this text will provide a useful guide into the field, enabling NMR spec-
troscopists and nonspecialists alike to evaluate critically the potentialities
and limitations of the method, and its applications in the primary literature.

KURT WÜTHRICH

Zürich, Switzerland
March 1986

v

Acknowledgments

The incentive to prepare this text for publication has been provided by my tenure of the George Fisher Baker Nonresident Lectureship in Chemistry at Cornell University in September and October 1983. I wish to express my sincere thanks to Professor Roald Hoffmann and his colleagues in the Department of Chemistry for the invitation to present the Baker lectures and for the friendship and hospitality extended to me and my family during our period of residence in Ithaca.

My greatest debt is to the many friends, colleagues, and students in my group and around the globe, who have with their own work and their criticism over so many years provided inspiration and new incentives for continued activity. I would in particular like to mention the privilege of long standing, fruitful, and enjoyable collaboration with Professor R. R. Ernst and Professor G. Wagner, who also read large portions of this manuscript. Generous support in the preparation of this book was provided by my collaborators Dr. W. Braun, Dr. W. Chazin, Dr. T. Havel, Dr. A. Kline, Dr. W. Leupin, Dr. N. Müller, and Dr. E. Wörgötter and my students M. Billeter, G. Otting, T. Schaumann, P. Schultze, and H. Widmer, who gave permission to include unpublished results from their ongoing projects, and carefully read and made useful comments on various portions of the manuscript.

I am most grateful to those authors, publishers, and learned societies who gave permission to reproduce illustrations for which they hold the copyright. Due acknowledgment is made in each case in the text.

Finally I wish to thank Mrs. E. Huber and Mrs. E. H. Hunziker-Kwik, for having accepted the difficult task of typing the manuscript from my handwritten notes, Mr. R. Marani for the preparation of the reference list, and Mrs. E. H. Hunziker-Kwik for the carefully prepared illustrations.

K.W.

Contents

List of Tables

Symbols and Abbreviations

Expressions, symbols, and units commonly used in nuclear magnetic resonance are introduced in Section 1.4 and at the outset of Chapter 5 [comprehensive listings may be found, e.g., in Ernst et al. (1986) and Jardetzky and Roberts (1981)]. The following are less common abbreviations:

NOE	Nuclear Overhauser enhancement or nuclear Overhauser effect
TOE	Truncated driven NOE (a one-dimensional NMR experiment for measurement of NOE's)
2D NMR	Two-dimensional nuclear magnetic resonance
COSY	2D correlated spectroscopy
SECSY	2D spin echo correlated spectroscopy
FOCSY	2D foldover-corrected spectroscopy
NOESY	2D nuclear Overhauser and exchange spectroscopy
2D-J spectroscopy	2D J-resolved spectroscopy
MQ, ZQ, 2Q	Multiple quantum, zero quantum, two quantum (e.g., MQ spectroscopy)
MQF, 2QF	Multiple quantum filter, two quantum filter (e.g., 2QF-COSY)

All word combinations ending with COSY describe experiments for studies of scalar (through-bond) spin–spin couplings, for example,

RELAYED-COSY	Relayed coherence transfer spectroscopy

All word combinations ending with NOESY describe experiments for studies of dipolar (through-space) couplings manifested in nuclear Overhauser effects (or for studies of exchange phenomena), for example,

ω_1-decoupled NOESY   NOESY with homonuclear broadband decoupling along the ω_1 frequency axis

For the amino acids the IUPAC–IUB one-letter or three-letter symbols are used (Table 2.2). Individual atoms are identified using a prefix (e.g., NH is a backbone amide proton, 2H a hydrogen atom bound to the aromatic carbon atom 2, moreover αH, βH, βCH$_2$, γCH$_3$, 1NH, δNH$_2$). For the interpretation of certain NMR measurements, *pseudostructures* for amino acids are used, which contain the *pseudoatoms* K, L, M, P, Q, and QR as specified in Table 10.2. Polypeptide conformations are described by the torsion angles ϕ, ψ, ω, and χ^j (Fig. 7.2). The following abbreviations are used for proteins:

BPTI	Basic pancreatic trypsin inhibitor from bovine organs
cBPTI	Cyclic analog of BPTI
Inhibitor K	Trypsin inhibitor homologue K from the venom of *Dendroaspis polylepis polylepis*
BUSI	Proteinase inhibitor IIA from bull seminal plasma
Tendamistat	αAmylase polypeptide inhibitor from *Streptomyces tendae*
lac headpiece	DNA-binding domain 1–51 of the *lac* repressor from *E. coli*
metallothionein-2	Metallothionein isoprotein 2 from rabbit liver
micelle-bound glucagon	Glucagon bound to perdeuterated dodecylphosphocholine micelles.

For nucleosides or nucleotides the IUPAC–IUB one-letter symbols are used (Table 2.5). Individual atoms are identified using a prefix (Table 2.5, e.g., 1'H, 2H, 3NH, 6NH$_2$). Polynucleotide conformations are described by the torsion angles α–ζ, ν_0–ν_4, and χ (Fig. 11.2).

Spin systems of the protons in amino acid residues and nucleotides are described by upper case letters (e.g., AX, AMX, A$_3$B$_3$MX; see Tables 2.2 and 2.5, Sections 2.1 and 2.2). The following symbols are used to identify types of spin systems representing different groups of amino acids:

□	AMX of αCH–βCH$_2$ in Cys, Ser, Asp, Asn, Phe, Tyr, His, Trp
■	Pro, Met, Glu, Gln, Arg, Lys
◩	Cys, Asp, Asn
▲	Met, Glu, Gln
▼	Arg, Lys

Proton–proton distances have an important role in the analysis of the NMR data. In proteins, $d_{AB}(i,j)$ is the distance from proton A in amino acid residue i to proton B in residue j [e.g., $d_{\alpha N}(4,18)$, $d_{\alpha\beta}(i,i+3)$; $d_{\alpha N}(i,i+1) \equiv d_{\alpha N}$; see Section 7.1]. In nucleic acids the distances $d_i(A;B)$, $d_s(A;B)$, $d_{pi}(A;B)$, and $d_{ps}(A;B)$ are used as specified in Section 11.1.

The expressions CONFOR, DISGEO, and DISMAN indicate computer programs for the structural analysis of NMR data. The RMSD is the *root mean square distance* between different conformers.

THE GEORGE FISHER BAKER
NON-RESIDENT LECTURESHIP
IN CHEMISTRY AT
CORNELL UNIVERSITY

NMR of Proteins
and Nucleic Acids

CHAPTER 1
Introduction and Survey

The first nuclear magnetic resonance (NMR) experiments with biopolymers were described over 30 years ago, and the potentialities of the method for studies of structure and dynamics of proteins and nucleic acids were long anticipated. However, in practice, initial progress was slow because of limitations imposed both by the available instrumentation and by the lack of suitable samples of biological macromolecules. Nuclear magnetic resonance of proteins and nucleic acids is represented in the literature by less than 30 papers during the years up to 1965, approximately 200 papers up to 1970, 4000 papers up to 1980, and of the order 500 to 1000 papers a year from 1981 onward. The work up to 1980 was covered in several textbooks (Dwek, 1973; Govil and Hosur, 1982; James, 1975; Jardetzky and Roberts, 1981; Wüthrich, 1976) and collections of review articles (e.g., Dwek et al., 1977; Opella and Lu, 1979; Shulman, 1979), and most papers published during the early 1980s used experimental procedures that are covered in these texts. This monograph describes a new approach for NMR studies of structure, dynamics, and intermolecular interactions of proteins and nucleic acids, which was developed since approximately 1977. It relies on sequence-specific resonance assignments obtained entirely from NMR experiments and knowledge of the chemical structure. The practicability of this approach was greatly aided by the fact that adaptation of two-dimensional (2D) NMR experiments for studies of biomacromolecules was started very shortly after the initial publications on 2D NMR (Aue et al., 1976a,b; Nagayama et al., 1977).

1.1. CONFORMATIONAL STUDIES BY NMR AND BY DIFFRACTION METHODS

Since the pioneering work of Perutz and Kendrew in the late 1950s, single crystal X-ray studies have set standards for protein and nucleic acid spatial structure determination and provided data on over 200 proteins, several tRNA's, and a selection of synthetic oligonucleotides (Richardson, 1981; Saenger, 1984; Schulz and Schirmer, 1979). The NMR approach described in this book is complimentary to X-ray crystallography in different ways:

1. The NMR studies use noncrystalline samples (e.g., solutions in aqueous or nonaqueous solvents, or detergent-solubilized biopolymers in mixed micelles). If NMR assignments and spatial structure determination by NMR can be obtained without reference to a corresponding crystal structure, a meaningful comparison of the conformations in single crystals and in noncrystalline states can be obtained.

2. Nuclear magnetic resonance can be applied to molecules for which no single crystals are available. Thus new structures can be obtained, which are not available from X-ray studies.

3. The solution conditions for NMR studies (e.g., pH, temperature, ionic strength, buffers) can usually be varied over a wide range. This opens the possibility for comparative studies under native and denaturing solution conditions, or for investigations of intermolecular interactions with other solute molecules.

4. For a characterization of the internal dynamics of biomacromolecular structures, NMR provides direct, quantitative measurements of the frequencies of certain high activation energy motional processes and at least semiquantitative information on additional high frequency processes. In comparison, X-ray structure determinations may include an outline of the conformation space covered by high frequency structural fluctuations. Furthermore, neutron diffraction in single crystals and NMR in solution can both be employed for studies of the exchange of labile protons, thus potentially enabling direct comparison of the molecular dynamics in the different states.

1.2. THE PIVOTAL ROLE OF SEQUENCE-SPECIFIC NMR ASSIGNMENTS

Since approximately 1981 methods have become available for obtaining nearly complete, sequence-specific resonance assignments in biopolymers. ^{1}H NMR assignments have been described to date for more than 10 small proteins, a selection of synthetic DNA fragments and, to a more limited extent, for tRNA's and fragments of rRNA and mRNA. Only a fraction of the NMR data are needed for obtaining the resonance assignments, and once these are available the information content of the remaining data is decisively increased.

As a first illustration we consider the structural analysis of intramolecular nuclear Overhauser effects (NOE) in a biopolymer chain. A NOE between two hydrogen atoms (or groups of hydrogen atoms) is observed if these hydrogens are located at a shorter distance than approximately 5.0 Å from each other in the molecular structure. Typically, a large number of NOE's are observed in globular proteins or in double-helical nucleic acids, indicating that these molecular structures contain numerous pairs of closely spaced hydrogen atoms (upper part of Fig. 1.1). Combined with resonance assignments these distance constraints can be attributed to specified sites along the polymer chain and hence the NOE experiments define the formation of loops by chain segments of variable lengths (Fig. 1.1). With model building or the use of suitable mathematical procedures, notably distance geometry (Blumenthal, 1970), the conformation space can then be searched for spatial arrangements of the polymer chain that are compatible with the experimental distance constraints. Sequence-specific resonance assignments thus provide a basis for systematic procedures toward spatial structure determination of noncrystalline biopolymers (Wüthrich et al., 1982).

As a second example we consider investigations of intermolecular interactions with biopolymers. Nuclear Overhauser effects can manifest short distances between nuclear spins located in the different, interacting molecules, and additional data on close intermolecular contacts in the complexes formed may be obtained from other experiments. Without sequence-specific resonance assignments such data can merely indicate that complex formation has occurred, but when combined with resonance assignments they identify the sites of intermolecular contacts in the polymer chain. As is

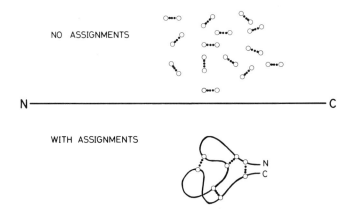

Figure 1.1. Information content of ¹H–¹H NOE's in a polypeptide chain with and without sequence-specific resonance assignments. Open circles represent hydrogen atoms of the polypeptide. The polypeptide chain is represented by the horizontal line in the center.

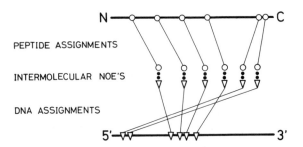

Figure 1.2. Identification of contacts between a polypeptide and a polynucleotide by intermolecular NOE's and sequence-specific NMR assignments. Circles and triangles represent nuclear spins in the polypeptide and the polynucleotide, respectively.

schematically shown in Figure 1.2 for a polypeptide chain interacting with a polynucleotide chain, such studies can initially be conducted on the primary structure level. They may then be combined with the results from investigations on the solution conformations of the interacting species. Besides studies of protein–nucleic acid interactions, potential applications of these principles include identification of the biopolymer contact sites in complexes

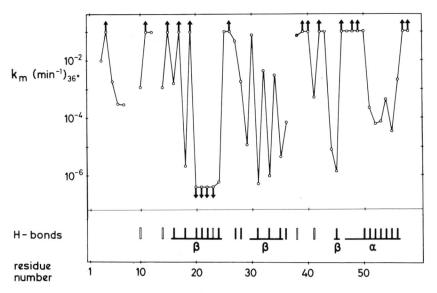

Figure 1.3. Backbone amide-proton exchange rates k_m in BPTI plotted versus the amino acid sequence. Near the bottom the sequence locations of α and β secondary structures and of hydrogen-bonded amide protons are indicated (from Wagner and Wüthrich, 1982b).

with small molecules, for example, in interactions of DNA with drugs, or interactions of proteins with effector molecules.

The use of NMR for studies of molecular dynamics relies on the observation of spectral properties in distinct NMR lines that can be correlated with intramolecular motional processes. Once the NMR lines used for this purpose have independently been assigned, the intramolecular motions can be attributed to specified locations in the molecular structure, thus providing a map of the internal motility across the molecule. As an illustration Figure 1.3 shows a plot versus the amino acid sequence of the individual backbone amide-proton exchange rates in the small protein basic pancreatic trypsin inhibitor (BPTI). In this presentation the dynamic processes are mapped out on the level of the primary structure. If the molecular conformation is known, a corresponding map can be produced for the spatial structure. In further discussions of Figure 1.3 in Chapters 9 and 14 we shall see that these proton exchange data manifest the entire hydrogen-bonding network of the secondary structure elements in BPTI and contain additional information on time fluctuations of the molecular structure.

1.3. ABOUT THIS BOOK

This text deals with the methods used for obtaining sequence-specific NMR assignments in proteins and nucleic acids, and with the analysis of biopolymer NMR data on the basis of such resonance assignments. The selection of the material covered was guided by its relevancy for this main theme. Section 1.4 is a brief survey of NMR and the principal NMR parameters, which was included for readers who may be newcomers to the NMR field. Part I describes fundamental aspects of biopolymer NMR spectra and presents an introduction to 2D NMR techniques. Parts II and III are the heart of the book. They treat the specific approaches for resonance assignments and spatial structure determination in proteins and nucleic acids. Part IV is a brief discussion relating to the biological significance of the results on biopolymer structure, dynamics, and intermolecular interactions that may be obtained with these NMR techniques.

The book is written at a time when the field covered is still in rapid progress, characteristic of an early stage of development. The author expects in particular that NMR studies of biopolymers will continue to benefit from advances in NMR instrumentation and design of new experiments, and that they will be largely influenced by the use of more powerful computational methods for handling and analysis of NMR spectra and for the structural interpretation of NMR data. Given this situation, the main emphasis of the text is on a nonmathematical description of fundamental principles, and on formulating questions to be answered by present and future technical procedures.

1.4.   NMR AND NMR PARAMETERS

The purpose of this section is a brief survey of the principal NMR parameters which, on the one hand, determine the appearance of the NMR spectra and, on the other hand, contain information on the molecules under investigation. As a reference for the definition of these parameters we start with a description of some elementary features of NMR experiments.

In a high resolution NMR experiment a glass tube containing a solution of the molecule of interest is placed in a *static magnetic field* B_0 and then subjected to irradiation by one or several *radio-frequency (rf) fields*, B_1, B_2, B_3. Figure 1.4*A* presents a simple, classical picture of the effects of these fields on the *magnetization* of the sample under study. Under the influence of B_0 the nuclear spins are *polarized*, resulting in net macroscopic magnetization M in the direction of B_0. B_1 applied along the x' axis exerts a torque on M. A rf pulse of suitably chosen duration *(90° pulse)* thus rotates M into the x,y plane perpendicular to B_0. The *transverse magnetization* in the x,y plane precesses under the influence of the static magnetic field B_0 at the *resonance frequency* (or *Larmor frequency*) ν_0 and induces an electric current in a *detection coil*, which is the primary observation in a NMR experiment. [For work with biopolymers B_0 = 11.74 Tesla (T) is typically used. The corresponding Larmor frequency for ^{1}H is ν_0 = 500 Megahertz (MHz). The Larmor frequency is often given in units of radians/s, with $\omega_0(rad)$ = $2\pi\nu_0(Hz)$.] Since the system tends to return to the *thermodynamic equilibrium* situation with the magnetization oriented parallel to B_0, the transverse magnetization decays with time. The corresponding *free induction decay* (FID) is recorded during a period of approximately 1 s, and the NMR spectrum is then obtained by *Fourier transformation* (FT) of these data.

Figure 1.5 is well suited as an illustration for the following description of NMR parameters, since it includes also molecular rate processes. It shows the temperature dependence of the aromatic ^{1}H NMR lines of the four

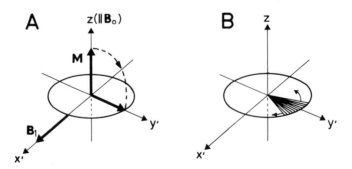

Figure 1.4. *(A)* Effects of external magnetic fields on a NMR sample. A rotating frame coordinate system is defined by x', y', and z, with z along B_0. *(B)* Decay of phase coherence during precession of the magnetization in the x,y plane.

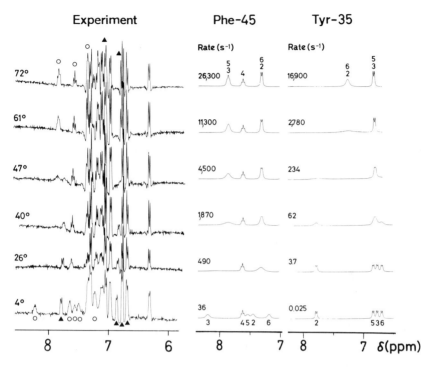

Figure 1.5. 1D ^{1}H NMR spectra at variable temperature of BPTI in D$_2$O (0.01 M, pD 7.8; 360 MHz). The region from 6.0 to 8.5 ppm is shown. The resonances of the aromatic rings of Phe-45 (○) and Tyr-35 (▲) were simulated in the center and on the right, respectively, using the ring flip rates indicated with the individual spectra (see text). The resonance assignments in the simulated spectra refer to the numeration in Figure 1.6 (from Wüthrich and Wagner, 1978).

phenylalanines and four tyrosines in BPTI. [This experiment and similar observations in different proteins had a considerable influence on present concepts of protein dynamics. It manifests 180° ring flips of Phe and Tyr (Fig. 1.6), with the flipping frequencies indicated next to the spectrum simulations in Figure 1.5, and thus presents direct evidence for structure fluctuations in the protein interior (arrows in Fig. 1.6). By now such ring flips are recognized as a common feature in globular proteins.]

1. The *chemical shift* (δ) defines the location of a NMR line along the rf axis. It is measured relative to a reference compound [for biopolymers usually a water soluble derivative of tetramethylsilane, e.g., *3-trimethylsilyl-propionate* (TSP), or *2,2-dimethyl-2-silapentane-5-sulfonate* (DSS)]. In frequency units the chemical shift is proportional to the applied static magnetic field, and therefore chemical shifts are customarily quoted in *parts per million* (ppm) units. Then, δ is primarily related to the chemical structure of the molecule studied. In the experimental spectrum at 4° (Fig. 1.5) the chemical

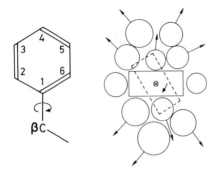

Figure 1.6. Tyr and Phe 180° ring flips about the βC–1C bond. The drawing on the right presents a view along the βC–1C bond of a flipping ring in the interior of a protein. The circles represent atom groups near the ring, and arrows indicate movements of atom groups during the ring flip.

shifts of the four ring protons of Tyr-35 and the five ring protons of Phe-45 are indicated by ▲ and ○, respectively. In NMR terminology, a resonance is at *high field* relative to another resonance if it has a smaller δ in ppm units.

2. *Spin–spin coupling constants (J)* characterize *scalar interactions* (through-bond) between nuclei linked via a small number of covalent bonds in a chemical structure. *J* is field independent and is customarily quoted in *hertz* (Hz). In Figure 1.5 spin–spin couplings are manifested by the partially resolved *fine structure* of the individual resonance lines. For example, for Tyr-35 at 4°, each of the four lines has a *doublet* fine structure.

3. The *NMR line intensity.* In a normal 1D NMR spectrum the relative intensities of different resonance lines reflect the number of nuclei manifested by these lines. For example, in the spectrum at 72° (Fig. 1.5), Phe-45 gives rise to two two-proton lines (3,5 and 2,6) and a one-proton line (4).

4. The *nuclear Overhauser enhancement* or *nuclear Overhauser effect* (NOE) is the fractional change in intensity of one NMR line when another resonance is irradiated in a *double irradiation experiment.* Nuclear Overhauser effects are due to *dipolar interactions* (through-space) between different nuclei and are correlated with the inverse sixth power of the internuclear distance. Nuclear Overhauser effects are customarily quoted in percent of the unperturbed resonance intensity.

5. The *longitudinal relaxation time* or *spin–lattice relaxation time* (T_1) describes the rate at which the magnetization **M** returns to the *thermodynamic equilibrium* orientation along $\mathbf{B}_0$ after a rf pulse (Fig. 1.4). T_1 is usually given in units of *seconds.* T_1 is correlated with the overall rotational tumbling of the molecule in solution and may be further affected by intramolecular mobility in flexible structures.

6. The *transverse relaxation time* or *spin–spin relaxation time* (T_2) describes the decay rate of the effective magnetization observed in the *x,y* plane after a 90° pulse (Fig. 1.4). T_2 is shorter than or equal to T_1. Differences between T_1 and T_2 arise because of loss of *phase coherence* among the individual spins that precess in the x,y plane and constitute the macroscopic, observable magnetization (Fig. 1.4B). T_2 is correlated with dynamic processes in the molecule under study; in particular it decreases monotonously with increasing molecular size, which presents a limiting factor for high resolution NMR with macromolecules.

7. The resonance *linewidth* in a NMR spectrum is of crucial practical importance, since it determines the *spectral resolution* that can be attained at a given field strength B_0. Assuming a *Lorentzian line shape,* the linewidth is customarily presented as the half-width at half-height of the line, $\Delta\nu_{1/2}$, in frequency units. In an ideal NMR experiment with a kinetically stable system, the linewidth is simply related to T_2 (i.e., $\Delta\nu_{1/2} = 1/2\pi T_2$), but it can be dramatically affected by exchange processes and by experimental artifacts. Figure 1.5 presents an example of *exchange broadening* of NMR lines. All the spectral simulations in this figure used a linewidth, $\Delta\nu_{1/2}$, of 2.5 Hz. At 4° the line shapes for Tyr-35 manifest this linewidth, since at this temperature the exchange processes due to flipping motions of this ring (Figure 1.6) are too slow to be noticed. As a consequence of higher flip frequencies (Fig. 1.5), the lines of Phe-45 at 4°C are exchange broadened, and in some spectra at higher temperatures individual lines of both residues are broadened beyond detection. Experimental artifacts include, in particular, effects from *inhomogeneity of the static field* B_0. This can be readily appreciated from Figure 1.4. In the x,y plane the spins precess under the influence of B_0. If B_0 is inhomogeneous over the sample volume, individual spins in the macroscopic sample will precess at somewhat different frequencies, resulting in loss of *phase coherence* (Fig. 1.4B). Such artifactual contributions to the experimentally observed, apparent T_2 relaxation times, T_2^*, are customarily denoted by T_2^+, with $1/T_2^* = 1/T_2 + 1/T_2^+$.

8. *NMR time scales.* A qualitative appreciation of NMR time scales is indispensable for designing NMR experiments and analyzing NMR spectra, and further provides a survey of the *dynamic processes* that may be studied by NMR. The time scales accessible in a given experiment are largely determined by the equipment used, notably by the field strength B_0. A fundamental NMR time scale is defined by the *resonance frequency* ν_0 (e.g., $\nu_0 = 500$ MHz for ^1H at $B_0 = 11.74T$). Molecular rate processes with characteristic times in the range from approximately 10^{-7} to 10^{-12} s are thus directly manifested in the relaxation parameters T_1 and T_2. (This includes the overall rotational tumbling of the molecules in solution as well as segmental intramolecular motions.) Much slower NMR time scales result from reversible exchange of nuclei between two or several different environments, either by *chemical exchange* or by *physical, dynamic processes.* As an illustration we consider the spectrum of Tyr-35 in Figure 1.5. There are two pairs of protons (2,6 and 3,5) that exchange locations in the protein structure as a result of each 180° ring flip (Fig. 1.6). The two locations between which this exchange occurs are characterized by different chemical shifts. At 360 MHz the chemical shift differences in frequency units, $\Delta\delta(Hz)$, are 382 Hz for the 2,6 protons and 14 Hz for the 3,5 protons. Three different regimes can be characterized by the ratio of the ring flip frequency k and the chemical shift difference $\Delta\delta(Hz)$. $k \ll 2\pi\Delta\delta$ describes *slow exchange* on the NMR time scale, $k \approx 2\pi\Delta\delta$ is *intermediate exchange*, and $k \gg 2\pi\Delta\delta$ is *fast exchange*. In the slow exchange situation, separate, narrow resonances are observed for the individual protons (spectra at 4 and 26°). In fast exchange a single, narrow resonance of intensity corresponding to two protons is observed at the average of the chemical shifts of the two contributing protons (3,5-proton line at 61 and 72°). Intermediate exchange is characterized by a continuous transition between the two limiting situations, with pronounced exchange broadening of the lines (2,6-proton resonances between 40 and 72°, 3,5-proton resonances at 40 and 47°). An important message is that a particular NMR experiment may cover largely different time scales, depending on the chemical shift differences between the exchanging nuclei. Additional NMR time scales can be defined for exchange averaging of spin–spin couplings or relaxation times.

PART I
THE FOUNDATIONS: STRUCTURE AND NMR OF BIOPOLYMERS

CHAPTER 2
NMR of Amino Acid Residues and Mononucleotides

The monomeric building blocks of biopolymers were exhaustively investigated by NMR (Bovey, 1972; Jardetzky and Roberts, 1981; Wüthrich, 1976). This chapter surveys those features of amino acid residues and nucleotides that are needed as reference material for NMR studies of proteins and nucleic acids.

Table 2.1 lists important parameters of the four isotopes with spin $\frac{1}{2}$ that are ubiquitous in biopolymers and can be used for high resolution NMR experiments (Pople et al., 1959). Most of the work discussed in this book relies on observation of ^{1}H resonances, thus exploiting the outstanding properties of this isotope. First, for ^{1}H the relative and absolute sensitivity and hence the ease of observation by NMR are higher than for the other isotopes. Second, because of the peripheral locations of hydrogen atoms in the molecular structures, ^{1}H NMR signals have a pronounced sensitivity to nonbonding interactions, such as those between different molecular segments in globular proteins or between the individual strands in nucleic acid duplexes. For ^{13}C and ^{15}N the low natural abundance contributes to the low sensitivity for observation of these nuclei, but it presents also the opportunity for site-specific NMR observations using isotope labels. Similarly, observation of ^{31}P may be employed for NMR studies of nucleic acids in complexes with phosphorous-free compounds, without spectral interference by these other components. In special situations additional isotopes that are not commonly found in natural biopolymers may be used for NMR experiments. Examples are diamagnetic metal ions (e.g., ^{111}Cd, ^{113}Cd, and ^{199}Hg) in metalloproteins, or ^{19}F in compounds that were modified by chemical or biochemical methods for the NMR investigations.

TABLE 2.1. Nuclear Properties of Selected Isotopes

Isotope ($I=1/2$)	$\gamma \times 10^{-7}$ (rad $T^{-1}s^{-1}$)	ν at 11.74T (MHz)	Natural Abundance (%)	Sensitivity Rel.[a]	Sensitivity Abs.[b]
^1H	26.75	500.0	99.98	1.00	1.00
^{13}C	6.73	125.7	1.11	1.6×10^{-2}	1.8×10^{-4}
^{15}N	-2.71	50.7	0.37	1.0×10^{-3}	3.8×10^{-6}
^{31}P	10.83	202.4	100	6.6×10^{-2}	6.6×10^{-2}

[a] Relative sensitivity at constant field for equal number of nuclei.
[b] Product of relative sensitivity and natural abundance.

2.1. SPIN SYSTEMS IN AMINO ACID RESIDUES

The chemical shifts for the common amino acid residues (Fig. 2.1, Table 2.2) in "random coil" polypeptides (Table 2.3) are a reference both for a qualitative understanding of the general features of a protein spectrum and for certain quantitative analyses of the latter. To afford an easily accessible general view of the distribution of *random coil ^1H chemical shifts*, the hydrogen atoms with similar shifts were grouped together in the presentation of Table 2.4.

The data in Table 2.3 were obtained from NMR measurements in aqueous solution of 20 tetrapeptides H-Gly-Gly-Xxx-Ala-OH, where in each peptide Xxx was a different one of the 20 common amino acids (Bundi and Wüthrich, 1979a). Because the influence of the chain terminal backbone amino and carboxyl groups is to a good approximation limited to the spectral properties of the terminal residues, these model peptides mimic rather faithfully the behavior of nonterminal residues in random coil polypeptides. pH titration effects are essentially limited to the two terminal residues and the interior residues with ionizable side chains. ^1H chemical shifts for the protonated and deprotonated species and the corresponding intrinsic titration shifts for the common amino acid residues were tabulated by Bundi and Wüthrich, (1979a,b). (Among the early NMR studies of proteins, observation of pH dependent chemical shifts produced some of the most interesting

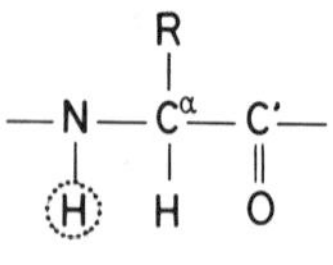

Figure 2.1. Amino acid residue. R is the side chain (see Table 2.2). The dotted circle identifies the labile, but often NMR-observable amide proton.

TABLE 2.2. Side Chains R (see Fig. 2.1) and Three-Letter and One-Letter Symbols for the 20 Common Amino Acids, and Spin Systems of the Nonlabile Hydrogen Atoms in the Molecular Fragments H–αC–R[a,b]

H	CH_3	$\overset{O}{\underset{CH_2}{|}}$	$\overset{S}{\underset{CH_2}{|}}$
Gly, G	Ala, A	Ser, S	Cys, C
AX	A_3X	AMX	AMX

$\underset{C-H}{CH_3\ CH_3}$	$\underset{C-H}{O\quad CH_3}$	$\underset{CH_2}{O=C-O}$	$\underset{CH_2}{O=C-NH_2}$
Val, V	Thr, T	Asp, D	Asn, N
A_3B_3MX	A_3MX	AMX	AMX

$\underset{CH_2}{\overset{CH_3\ CH_3}{C-H}}$	$\begin{array}{c}CH_3\\CH_2\\H-C-CH_3\end{array}$	$\begin{array}{c}\overset{+}{NH_3}\\CH_2\\CH_2\\CH_2\\CH_2\end{array}$	$\begin{array}{c}NH_2\ \ NH_2\\ \overset{+}{C}\\N-H\\CH_2\\CH_2\\CH_2\end{array}$
Leu, L	Ile, I	Lys, K	Arg, R
A_3B_3MPTX	$A_3MPT(B_3)X$	$A_2(F_2T_2)MPX^c$	$A_2(T_2)MPX^c$

$\begin{array}{c}O=C-O\\CH_2\\CH_2\end{array}$	$\begin{array}{c}O=C-NH_2\\CH_2\\CH_2\end{array}$	$\begin{array}{c}CH_3\\S\\CH_2\\CH_2\end{array}$	$\begin{array}{c}CH_2\\CH_2\quad CH_2\\-N\quad\quad \alpha CH-\end{array}$
Glu, E	Gln, Q	Met, M	Pro, P[d]
AM(PT)X	AM(PT)X	$AM(PT)X+A_3$	$A_2(T_2)MPX^c$

TABLE 2.2.　(Continued)

His, H	Phe, F	Tyr, Y	Trp, W
AMX+AX^e	AMX+	AMX+AA'XX'	AMX+A(X)MP+
	AMM'XX'		A

[a] Also indicated is the numeration used for the aromatic ring atoms. Not indicated is the standard identification of the other side chain heavy atoms by lower case Greek letters, with βC being next to αC (IUPAC–IUB Commission on Biochemical Nomenclature, 1970).

[b] Labile protons that can under certain conditions be observed by NMR in aqueous solution are shown in dotted circles. Those labile protons, which are not usually observed, are indicated by small filled circles.

[c] For simplicity the spin systems for Arg, Lys, and Pro were written with the assumption that with the exception of the β position, each methylene group gives rise to a single two-proton resonance.

[d] The structure for Pro includes the backbone atoms αCH and N.

[e] In His, 2H and 4H appear often as two singlet lines, but the connectivity through the small four-bond coupling of approximately 1 Hz was observed in several proteins. 2H can be exchanged with deuterium of the solvent, D_2O, within a period of several hours to several months.

and reliable results; for reviews see, e.g., Jardetzky and Roberts, 1981; Markley, 1979.)

Listings of chemical shifts and pH titration effects obtained from systematic studies of amino acid residues in model peptides are also available for ^{13}C NMR in aqueous solution (Keim et al., 1973a,b; Richarz and Wüthrich, 1978a), ^1H NMR in dimethyl sulfoxide (Bundi et al., 1975), and ^{13}C NMR in dimethyl sulfoxide (Grathwohl and Wüthrich, 1974).

In D_2O solution the labile protons (Fig. 2.1, Table 2.2) can be replaced by deuterium, so that the ^1H NMR spectrum contains only the resonance lines of the carbon-bound hydrogen atoms. For each amino acid residue these nonlabile protons constitute one or several *spin systems*. A spin system is a group of spins that are connected by scalar (through-bond) spin–spin couplings *J*. For practical purposes, in biological macromolecules such *J* couplings can usually be observed between hydrogen atoms that are separated by three or less covalent bonds. Figure 2.1 and Table 2.2 then show that the nonlabile protons of all but five amino acid residues form a single spin system. The exceptions are Met, where ϵCH_3 is not connected with the

TABLE 2.3. Random Coil ¹H Chemical Shifts for the 20 Common Amino Acid Residues[a]

Residue	NH	αH	βH	Others
Gly	8.39	3.97		
Ala	8.25	4.35	1.39	
Val	8.44	4.18	2.13	γCH_3 0.97, 0.94
Ile	8.19	4.23	1.90	γCH_2 1.48, 1.19
				γCH_3 0.95
				δCH_3 0.89
Leu	8.42	4.38	1.65,1.65	γH 1.64
				δCH_3 0.94, 0.90
Pro [b]		4.44	2.28,2.02	γCH_2 2.03, 2.03
				δCH_2 3.68, 3.65
Ser	8.38	4.50	3.88,3.88	
Thr	8.24	4.35	4.22	γCH_3 1.23
Asp	8.41	4.76	2.84,2.75	
Glu	8.37	4.29	2.09,1.97	γCH_2 2.31, 2.28
Lys	8.41	4.36	1.85,1.76	γCH_2 1.45, 1.45
				δCH_2 1.70, 1.70
				ϵCH_2 3.02, 3.02
				ϵNH_3^+ 7.52
Arg	8.27	4.38	1.89,1.79	γCH_2 1.70, 1.70
				δCH_2 3.32, 3.32
				NH 7.17, 6.62
Asn	8.75	4.75	2.83,2.75	γNH_2 7.59, 6.91
Gln	8.41	4.37	2.13,2.01	γCH_2 2.38, 2.38
				δNH_2 6.87, 7.59
Met	8.42	4.52	2.15,2.01	γCH_2 2.64, 2.64
				ϵCH_3 2.13
Cys	8.31	4.69	3.28,2.96	
Trp	8.09	4.70	3.32,3.19	2H 7.24
				4H 7.65
				5H 7.17
				6H 7.24
				7H 7.50
				NH 10.22
Phe	8.23	4.66	3.22,2.99	2,6H 7.30
				3,5H 7.39
				4H 7.34
Tyr	8.18	4.60	3.13,2.92	2,6H 7.15
				3,5H 6.86
His	8.41	4.63	3.26,3.20	2H 8.12
				4H 7.14

[a] Data for the nonterminal residues X in tetrapeptides GGXA, pH 7.0, 35°C [from Bundi and Wüthrich (1979a), except that more precise data were obtained for Leu, Pro, Lys, Arg, Met, and Phe using new measurements at 500 MHz].
[b] Data for *trans*-Pro.

TABLE 2.4. Groups of Hydrogen Atoms in the Common Amino Acid Residues with Similar Random Coil ¹H Chemical Shifts[a]

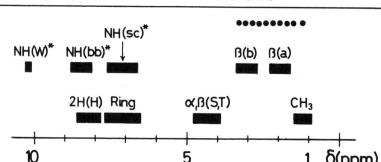

Code	δ (ppm)	Comments
CH$_3$	0.9-1.4	
β(a)	1.6-2.3	βH of V,I,L,E,Q,M,P,R,K
β(b)	2.7-3.3	βH of C,D,N,F,Y,H,W
●●●●	1.2-3.3	Other Aliphatic CH
α,β(S,T)	3.9-4.8	All αH, βH of S and T
Ring	6.5-7.7	Aromatic CH of F,Y,W; 4H of H
2H(H)	7.7-8.6	2H of H in the pH range 1-11
NH(sc)*	6.6-7.6	Side Chain NH of N,Q,K,R
NH(bb)*	8.1-8.8	Backbone NH
NH(W)*	10.2	Indole NH of W

[a] In model peptides the labile protons (identified by *) are only observed in H$_2$O solution. The singlet resonance of εCH$_3$ in Met is at 2.13 ppm (Table 2.3).

αCH–βCH$_2$–γCH$_2$ fragment, and the four aromatic residues, where the ring protons are not connected with the αCH–βCH$_2$ fragments.

To describe the spin systems, we use a notation adapted from Pople et al. (1959). Spins with different chemical shifts are denoted by different upper case letters, with the letter representing the highest field resonance preceding the others in the alphabet. Neighboring letters in the alphabet indicate nuclei with strong coupling (i.e., the chemical shift difference in frequency units is comparable to J), nonneighboring letters indicate weak coupling. A group of n equivalent spins with identical chemical shifts and identical spin–

spin couplings to all other spins in the system are denoted A_n. The two protons of Gly can thus form either an AX, AB, or A_2 spin system, depending on the relative chemical shifts. Table 2.2 includes a description of the ^{1}H spin systems for the common amino acids. Weak coupling was arbitrarily assumed throughout (in practice strong coupling prevails among most side chain methylene protons and in the Phe ring). The proton that resonates at highest field (Table 2.3) is always listed first and the spins are ordered in the direction from the periphery toward αH. Branching of the spin–spin coupling network away from this direction is indicated by parentheses (e.g., in Lys, A_2 corresponds to γCH_2, which is coupled to δCH_2–εCH_2, represented by F_2T_2, and to βCH_2–αCH, represented by MPX). The rings of Phe and Tyr contain two pairs of symmetry-related protons, 3H and 5H, and 2H and 6H, respectively, which are chemical shift equivalent but have different couplings with other spins in the aromatic system. Such *chemical shift equivalence* of two spins is denoted by AA'.

Thinking in terms of the spin systems in Table 2.2 is conceptually helpful for the analysis of the complex ^{1}H NMR spectra of proteins. Using experiments capable of delineating ^{1}H–^{1}H J couplings, the NMR lines can be arranged in groups corresponding to complete spin systems. The number of resonance lines and the network of J connectivities in the spin systems reflect the symmetry of the amino acid covalent structure. These patterns are the only spectral properties of the amino acid residues that are strictly preserved in proteins (Chapter 3). Gly, Ala, Val, Leu, Ile, Thr, and Lys have unique spin systems among all amino acid residues, and similarly the four systems corresponding to the aromatic protons of Phe, Tyr, His, and Trp manifest the different symmetries of the molecular structures. There are 8 residues with AMX systems, 3 residues with AM(PT)X systems, and 2 residues with $A_2(T_2)$MPX systems (Table 2.2). For these 13 species additional information besides the symmetry of the ^{1}H spin system is needed for a unique identification by NMR. Finally, reference should at this point be made to Figures 1.5 and 1.6, where studies of the symmetry of the aromatic ^{1}H spin systems resulted in a characterization of the dynamic states of the rings of Phe and Tyr in a globular protein.

There are two obvious extensions to the spin systems in Table 2.2. First, the labile protons can be included, which increases the complexity of the spin systems for all amino acid residues except proline. Compared to working with the nonlabile protons alone, two additional residues, Arg and Pro, can then be uniquely identified from the topology of the ^{1}H spin system. Second, heteronuclear spin systems including ^{1}H, ^{13}C, and ^{15}N spins can be defined. Their exploitation has in practice been limited, since at natural isotope distribution the concentration of species containing a single ^{13}C or ^{15}N spin per amino acid residue corresponds to approximately one percent of the total concentration (Table 2.1), and the concentrations of species with multiple ^{13}C or ^{15}N spins are negligibly small.

2.2. SPIN SYSTEMS IN MONONUCLEOTIDES

Figure 2.2 shows a fragment from a natural polynucleotide chain and Table 2.5 presents the molecular structures of the building blocks in deoxyribonucleic acids (DNA) and ribonucleic acids (RNA). With studies similar to those with model peptides, reference NMR parameters were collected from oligonucleotides [e.g., Cheng and Sarma (1977) and Cheng et al. (1984) tabulated the ^{1}H and ^{31}P chemical shifts and the ^{1}H–^{1}H and ^{1}H–^{31}P coupling constants in all 16 naturally occurring deoxyribodinucleoside monophosphates; for ^{13}C chemical shifts and ^{13}C–^{31}P coupling constants see Lankhorst et al. (1984) and references therein]. However, it is quite impracticable to define "random coil" NMR parameters for the mononucleotide units in natural DNA or RNA that could generally serve as reference values for work with nucleic acids. Reasons for this are that the NMR chemical shifts show a pronounced sequence dependence even for "unstructured" flexible forms of the polynucleotides, and that measurements with low molecular weight model compounds may be strongly influenced by intermolecular interactions. Therefore, the ^{1}H NMR chemical shift ranges in Table 2.6 were chosen sufficiently wide so that they encompass the shifts observed in single stranded and duplex DNA or RNA fragments (e.g., Cheng and Sarma, 1977; Clore et al., 1984; Hare et al., 1983; Scheek et al., 1984). Table 2.6 can thus serve as an approximate but generally applicable guide for the analysis of nucleic acid ^{1}H NMR spectra.

In contrast to the situation in most amino acid residues, the spin systems of the nonlabile protons (Table 2.5) do not cover the complete mononucleotide structures, since there are no readily observable ^{1}H–^{1}H J couplings between the ribose moiety and the base. In Table 2.5 the ribose spin systems were written starting from 1'H and following the J-coupling network to 5″ H, whereby weak coupling was assumed throughout. In DNA the two pyrimidine bases C and T can unambiguously be identified from the unique spin systems, whereas the purine bases A and G give exclusively one-proton singlets, which cannot a priori be further distinguished on the basis of symmetry considerations. In RNA the two pyrimidine bases C and U also have identical spin systems.

Addition of the NMR-observable labile protons (Table 2.5) does not result

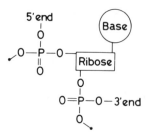

Figure 2.2. Nucleoside-3',5'-diphosphate (ribose and base structures in Table 2.5). The small filled circles indicate positions that may be occupied by protons at certain values of pH, but are not usually observable by ^{1}H NMR in aqueous media.

TABLE 2.5. Structures, Symbols, and ^{1}H Spin Systems of the β-D-Riboses and the Five Common Bases in DNA and RNA.[a]

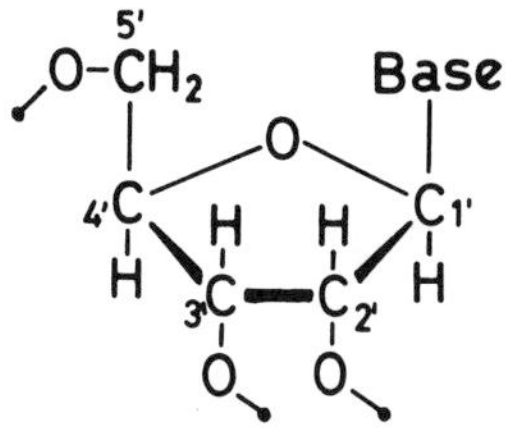

β-D-Ribose
XWTPMA

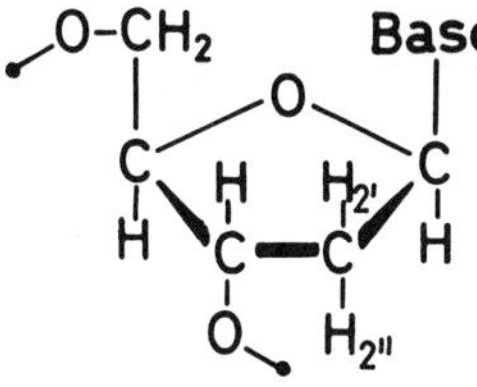

2'-Deoxy-β-D-Ribose
XAMWTNP

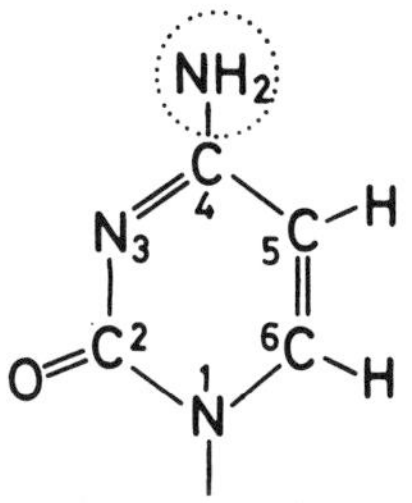

Cytosine, C
AX

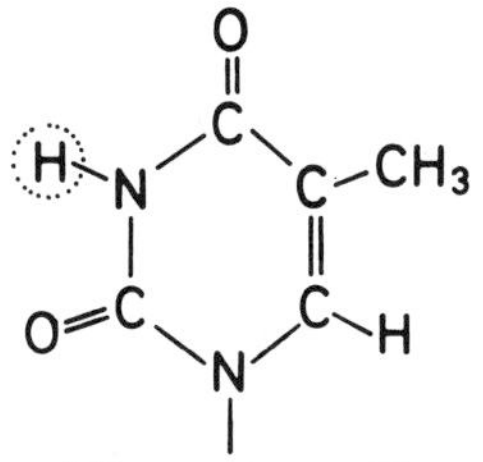

Thymine, T
A_3X

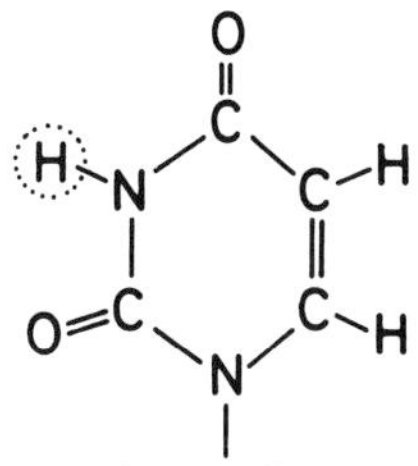

Uracil, U
AX

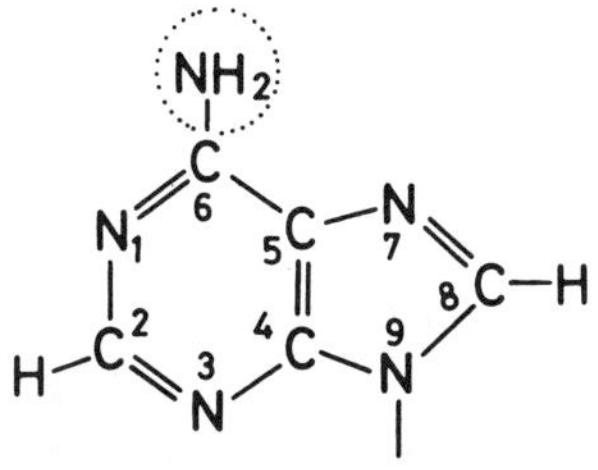

Adenine, A
A + A

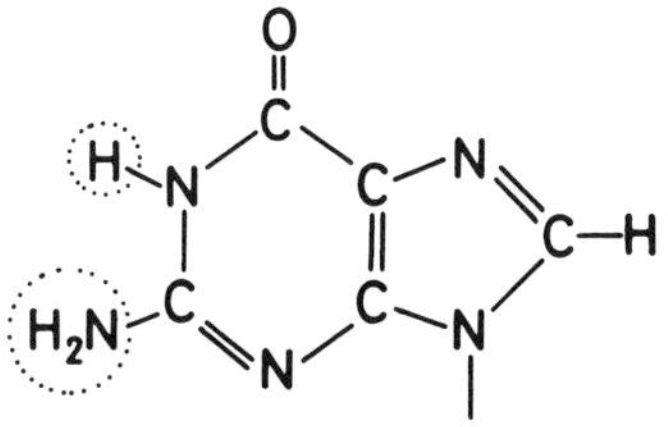

Guanine, G
A

[a] Indicated are the names for the structures, the one-letter symbols for the nucleotides, the standard numeration of atom positions, and the spin systems of the nonlabile hydrogen atoms. Labile protons that may be observed by NMR in H_2O solution are shown in dotted circles, those that cannot usually be observed are indicated by small filled circles. 8H in A and G can be exchanged with deuterium of the solvent, D_2O, within a period of several hours to several months.

TABLE 2.6. Approximate ^{1}H Chemical Shift Ranges in Single Stranded and Duplex DNA and RNA fragments[a]

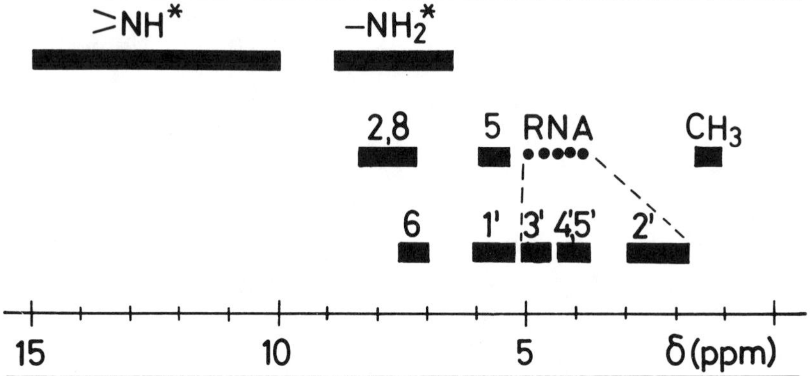

$>$NH* $-$NH$_2$*

2,8   5 RNA   CH$_3$

6   1' 3' 4,5' 2'

15   10   5   δ(ppm)

Code	δ (ppm)	Comments
2'	1.8–3.0	2'H, 2"H in DNA
4',5'	3.7–4.5	4'H, 5'H, 5"H in DNA
3'	4.4–5.2	3'H in DNA
•••••	3.7–5.2	2'H, 3'H, 4'H, 5'H, 5"H in RNA
1'	5.3–6.3	1'H
CH$_3$	1.2–1.6	CH$_3$ of T
5	5.3–6.0	5H of C and U
6	7.1–7.6	6H of C, T and U
2,8	7.3–8.4	8H of A and G, 2H of A
$-$NH$_2$*	6.6–9.0	NH$_2$ of A, C and G
$>$NH*	10 – 15	Ring NH of G, T and U

[a] The rather wide ranges result from the sequence effects on the shifts and for the labile protons (identified by *, observable only in H_2O) also from hydrogen bonding.

in expansions of the ^{1}H systems, since they are not coupled with the carbon-bound hydrogens. Therefore, the amino- and imino-proton resonances cannot in general be attributed to a unique base type on the basis of the spin system analysis.

In 3',5'-nucleoside diphosphates (Fig. 2.2) the ribose or deoxyribose protons 3'H, 5'H, and 5"H all have heteronuclear couplings to ^{31}P of the order 4–7 Hz, and the coupling 4'H–^{31}P is of the order 2–3 Hz (Cheng and Sarma, 1977). Unless these heteronuclear couplings are removed by ^{31}P broad band decoupling, a detailed analysis of the ribose ^{1}H NMR spectrum should therefore be based on consideration of the complete heteronuclear ^{1}H–^{31}P spin systems, which are present with 100% abundance (Table 2.1). Heteronuclear spin systems with ^{13}C and ^{15}N extend over the entire molecular structure of ribose and base, but at natural isotope distribution these are present only with very low abundance (Table 2.1).

2.3. INTRINSIC EXCHANGE RATES OF LABILE PROTONS

Labile protons in amino acid residues (Fig. 2.1, Table 2.2) and in pyrimidine and purine bases (Table 2.5) have important roles in NMR studies of proteins and nucleic acids. In addition to the NMR spectral parameters, the intrinsic rates of exchange with the solvent are of great interest as a reference for studies of macromolecular structure and dynamics.

Following the formalism of Eigen (1964), the exchange of labile protons is always acid or base catalyzed. The primary steps in this process are diffusion-controlled formation of a hydrogen-bonded complex with the catalyst, transfer of the labile proton to the catalyst in this hydrogen-bonded complex, and dissociation of the complex. Depending on the proton transfer equilibrium in the complex, this sequence of reaction steps may or may not produce a net transfer. The resulting forward rate for the proton transfer then becomes

$$k_{\mathrm{tr}} = k_{\mathrm{D}} \frac{10^{\Delta pK}}{10^{\Delta pK} + 1} \qquad (2.1)$$

k_{D} ($\approx 10^{10} \, M^{-1} \mathrm{s}^{-1}$ at 25°C) is the second-order rate constant for the diffusion controlled complex formation, and the second term describes the fraction of successful complexations, where ΔpK is the difference between the acidity constants of the proton donor and the proton acceptor in the complex. To obtain the first-order rate constant for the exchange, k_{tr} is multiplied with the concentration of the catalyst. Since there are in general several potential catalysts, an overall first-order rate constant taking account of exchange by all possible pathways can be written as

$$k_{\mathrm{intr}} = \sum_i k_{\mathrm{tr},i} \, [\mathrm{catalyst}]_i \qquad (2.2)$$

In aqueous solution OH^- and H_2O are potential base catalysts, and H_3O^+ and H_2O are potential acid catalysts. (Additional exchange pathways may result upon addition of pH buffers, in particular in the case of nucleic acids.) Using literature data for the pK_a's of the proton exchanging sites in amino acid residues and nucleotides (or in some cases related compounds), values for k_{intr} were estimated with Eq. (2.2) for H_2O solutions in the pH range 0–12 (Figs. 2.3 and 2.4).

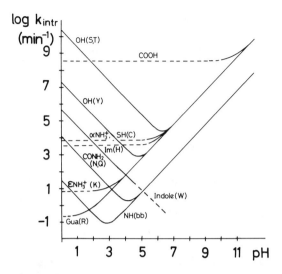

Figure 2.3. Logarithmic plots versus pH of approximate exchange rate constants k_{intr} computed with Eq. (2.2) for solvent accessible, labile protons of polypeptides in H_2O solution at 25°C. Broken lines represent lower limits for k_{intr} in situations where pK_a data were available either only for the base-catalyzed regime, or only for the acid catalysis. The individual curves are identified with the proton types and, where applicable, the residues types (Im stands for imidazole ring NH, Gua for guanidinium NH, bb for backbone) (adapted from Wüthrich and Wagner, 1979).

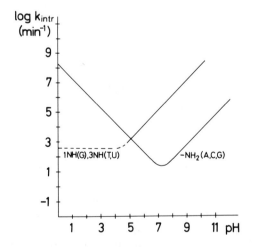

Figure 2.4. Logarithmic plots versus pH of approximate exchange rate constants k_{intr} for solvent accessible, labile base protons of polynucleotides in H_2O solution at 25°C. For all additional labile protons in polynucleotides (Fig. 2.2, Table 2.5), $k_{intr} \gtrsim 10^6$ min^{-1} over the entire pH range.

^{1}H NMR observation of labile protons is only feasible when the exchange with the solvent is slow on the NMR time scale. With the high field equipment used for studies of biopolymers, NMR observation will be possible if the exchange rate constant is smaller than approximately 1×10^3 min^{-1}. In practice, k_{intr} for backbone amide protons and the labile side chain protons of Arg, Lys, Asn, Gln, and Trp in polypeptides is sufficiently slow for NMR observation of these protons at slightly acidic pH and 25°C (Fig. 2.3). For all other labile protons in amino acid residues (Table 2.2) and all labile protons in nucleotides (Table 2.5) the intrinsic exchange rates are too fast over the entire pH range. In Tables 2.2 and 2.5 some of the protons with fast intrinsic exchange rates are shown in dotted circles, since they may exchange more slowly and hence be observable by NMR in the folded forms of proteins and nucleic acids (Chapter 3).

CHAPTER 3
NMR Spectra of Proteins and Nucleic Acids in Solution

Corresponding to the large size of the molecules, the NMR spectra of biopolymers are densely crowded with resonance lines. Obviously, these come from the nuclear spins contained in the constituent monomers (Figs. 2.1 and 2.2; Tables 2.2 and 2.5). However, there is no straightforward correlation between the NMR spectra of the low molecular weight components and those of the macromolecules, mainly for the following three reasons.

1. The spatial folding of the polymer chains is manifested in the chemical shifts, resulting in a dispersion of the shifts relative to random coil reference data from low molecular weight model compounds (e.g., Table 2.3). Because there is no strict periodicity either in biopolymer sequences or biopolymer conformations, multiple copies of identical monomeric units in the primary structure have in general different NMR chemical shifts. [This is in contrast to the situation encountered with synthetic polymers (Bovey, 1972).]

2. The exchange rates of labile protons in biopolymers can be several orders of magnitude slower than the intrinsic exchange rates (Figs. 2.3 and 2.4). As a consequence, these protons may be observed in the ^{1}H NMR spectra of the macromolecules under conditions where they could not be seen in small model compounds.

3. Different physical situations are encountered in NMR experiments with macromolecules or with small compounds. The slower diffusional motions of macromolecules in solution *(Brownian motion)* affect the spin relaxation (Fig. 3.1) and concomitantly the nuclear Overhauser enhancement (NOE) (Chapter 6) in fundamental ways. ^{1}H spin relaxation in diamagnetic molecules results primarily from ^{1}H–^{1}H dipole–dipole coupling modulated

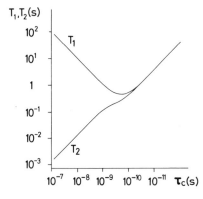

Figure 3.1. Logarithmic plots of the relaxation times T_1 and T_2 versus the correlation time τ_c. The curves have been computed for dipolar relaxation of two protons at a distance of 1.8 Å with a resonance frequency ν_0 of 500 MHz.

by the rotational tumbling of the vector joining the interacting spins (for a survey of relaxation mechanisms see Abragam, 1961; James, 1975; Jardetzky and Roberts, 1981; Wüthrich, 1976). These stochastic rotational motions are characterized by a *correlation time* τ_c, which is primarily dependent on the size of the molecules (τ_c for the overall rotational motions of spherical particles is proportional to the third power of the radius). For small molecules the frequencies of rotational tumbling τ_c^{-1} are high relative to the resonance frequencies ω_0 at presently available fields B_0, that is, $\omega_0\tau_c \ll 1$ ($\omega_0 = 2\pi\nu_0$; at 500 MHz, $\omega_0 - 3.1 \times 10^9$ s^{-1}). In this situation of *extreme motional narrowing*, $T_1 = T_2$ and both relaxation times increase proportional to τ_c^{-1} (Fig. 3.1). For macromolecules that rotate slowly relative to ω_0, so that $\omega_0\tau_c \gg 1$, T_1 increases proportionally to τ_c, and T_2 is proportional to τ_c^{-1}. As is illustrated in Figure 3.1, more complex τ_c dependences of T_1 and T_2 prevail in the intermediate range, where $\omega_0\tau_c \approx 1$.

Some general conclusions are valid for proteins and nucleic acids alike. First, because of the *chemical shift dispersion* each resonance line represents a single proton (or group of protons) in the macromolecule. Therefore, the molar concentration of the biopolymer solution must be equal to or even higher than the molar concentrations used for NMR studies of small molecules. Second, corresponding to the short relaxation times T_2 the NMR lines for macromolecules are broadened compared to small molecules. Further *line broadening* may result from aggregation or increase of the viscosity, which are not uncommon in concentrated solutions of biopolymers. On the other hand, line narrowing may result from intramolecular, segmental mobility. Third, to prevent saturation in macromolecules with slow T_1 relaxation (Fig. 3.1), long *recycle times* must be chosen at the expense of lowered *sensitivity*. Each of these three factors may in practice contribute to limitations on the size of the molecules for which particular NMR experiments can be applied.

The following sections discuss the ^{1}H NMR spectra of proteins, DNA fragments, and tRNA's in some more detail. Qualitatively similar principles would apply for the NMR spectra of the other nuclei in Table 2.1.

3.1. PROTEINS

In a pioneering paper McDonald and Phillips (1967) demonstrated that the amino acid side chain ^{1}H NMR lines in denatured, "random coil" polypeptide chains correspond closely to the sum of the resonances in the constituent amino acid residues, provided the increased macromolecular linewidths are taken into account. This observation can be rationalized by the assumption that all amino acid side chains in an extended, flexible polypeptide chain (Fig. 3.2A) are exposed to the same solvent environment, so that multiple copies of a specified amino acid in the sequence have identical chemical shifts. In Figure 3.3A a random coil spectrum of basic pancreatic trypsin inhibitor (BPTI) was thus computed as the sum of the resonances of the amino acid residues (Table 2.3) in the sequence of this protein. Increased complexity of the NMR spectrum of globular BPTI (Fig. 3.3B) results primarily from conformation-dependent chemical shift dispersion. Chemical shift dispersion arises because interior peptide segments in globular proteins are shielded from the solvent and are nearest neighbors to other peptide segments (Fig. 3.2B), so that different residues experience different microenvironments. In Figure 3.3 these effects are shown in detail for the Thr and Tyr residues in BPTI. Additional NMR lines between 7 and 11 ppm in the globular protein correspond to slowly exchanging amide protons (see Section 3.4).

Chemical shift dispersion by the spatial protein structure can also arise for protons within the same residue. This is exemplified in Figure 3.4 for a phenylalanine ring. In the covalent structure the two proton pairs 2,6 and 3,5 are in symmetry-related, equivalent positions. In an isotropic solvent this chemical shift equivalence is preserved, so that a spin system with symme-

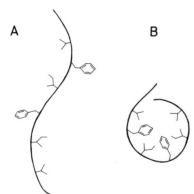

Figure 3.2. Scheme depicting the surroundings of amino acid side chains in an extended, "random coil" polypeptide chain (A) and in a globular protein (B).

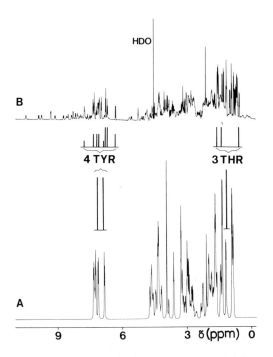

Figure 3.3. *(A)* Computed random coil NMR spectrum of nonlabile protons in BPTI. *(B)* ¹H NMR spectrum of a freshly prepared D_2O solution of BPTI (0.005 *M*, pD 4.5, 45°C; 360 MHz). Stick diagrams indicate the positions and intensities for γCH_3 of Thr-11, -32, and -54 and for the aromatic protons of Tyr-10, -21, -23, and -35 (from Wüthrich et al., 1982).

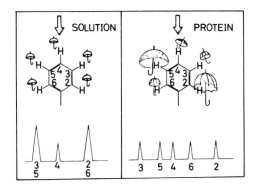

Figure 3.4. Nonbonding effects on the ¹H NMR spectrum of a phenylalanine ring in solution and in a globular protein. The size of the umbrellas reflects the shielding from B_0 (arrow) (adapted from Wüthrich and Wagner, 1978).

try AMM'XX' is observed. In the interior of a protein the twofold covalent structure symmetry is masked by different nonbonding *shielding effects* on the individual ring protons, so that the spin system symmetry is AGMPX. As discussed in Section 1.4 (Figs. 1.5 and 1.6), these effects from the protein environment can be averaged out by rapid rotational motions of the aromatic ring, whereby the extent of the conformation-dependent dispersion of the chemical shifts determines the NMR time scale for observation of these exchange processes.

At the origin of the chemical shift dispersion are anisotropic diamagnetic susceptibilities of molecular fragments in the polypeptide chain. Particularly large effects are due to the local *ring current fields* near aromatic amino acid residues (Fig. 3.5) and to intramolecular hydrogen bonds, but there is clear evidence that additional, smaller local fields are also manifested in the observed shifts (likely origins are, e.g., carbonyl groups and disulfide bonds). *Ring current shifts* arise because spins above or below the ring plane experience a local field that opposes B_0 (Fig. 3.5), and therefore the corresponding resonance lines are displaced to higher field (i.e., to the right in the spectrum). Conversely, spins located in the ring plane outside the confines of the ring are shifted to lower field. In BPTI the conformation-dependent shifts of the three Thr γCH_3 lines of -0.64, 0.16, and 0.38 ppm (Fig. 3.3) could be explained nearly quantitatively by the ring current effects predicted from the crystal structure, and the four one-proton lines located between 0.1 and 0.4 ppm were attributed to ring current shifted lines of proline β and γ protons (Perkins and Wüthrich, 1979). The influence of hydrogen bonds in BPTI on the amide- and α-proton chemical shifts was investigated on the basis of the sequence-specific NMR assignments. In the spectrum of Figure 3.3*B*, large downfield shifts of the α-proton lines between 5 and 6 ppm and the amide-

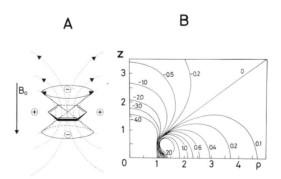

Figure 3.5. Ring current field of a benzene ring in solution. *(A)* Qualitative scheme. Inside the cone the local field is opposed to B_0, $\ominus$, outside it is parallel to B_0, $\oplus$. *(B)* Isoshielding lines (in parts per million) computed by Johnson and Bovey, 1959; z and ρ (in units of 1.39 Å) go through the ring center, with z perpendicular to the ring plane and ρ in the ring plane.

proton lines at the low field end of the spectrum were attributed to the formation of intramolecular hydrogen bonds (Wagner et al., 1983). Overall, studies of BPTI and other proteins, notably lysozyme and cytochrome c (e.g., Dalgarno et al., 1983; Perkins, 1982; Redfield et al., 1983) indicate that in diamagnetic proteins the conformation-dependent shifts of peripheral hydrogens in aliphatic side chains are nearly entirely due to ring current effects. For different types of protons, however, neither ring current effects nor hydrogen bonds appear to account for the observed shifts. For example, the large spread of the Tyr ring proton lines in BPTI (Fig. 3.3) is as yet unexplained (Redfield et al., 1983).

With regard to the analysis of the ^{1}H NMR spectra of new proteins, it is of interest to what extent the order of the random coil shifts in Table 2.4 is masked by conformation-dependent effects. This question can only be investigated in proteins for which ^{1}H NMR assignments are available. The presently available evidence suggests for all types of proteins that less than 10% of the resonance lines are shifted from their random coil class in Table 2.4 into the chemical shift ranges of other classes of protons, and that in diamagnetic proteins such mixing of proton types can always be associated with ring current shifts. (Low field shifts of backbone amide- and α-protons arising from hydrogen bonds may be large, but they do not usually cause overlap with chemical shift ranges for other proton types; see Table 2.4.) In this context it must be emphasized that in proteins containing Trp, aromatic prosthetic groups or paramagnetic centers, much larger conformation-dependent shifts can prevail than in BPTI. As an illustration, Figure 3.6 shows the spectrum of horse ferrocytochrome c, which contains numerous resonance lines in the region -4 to $+1$ ppm, with conformation shifts caused by the ring current field of the heme of up to -6 ppm. (The lines between 9 and 10 ppm correspond to the *meso* protons of the heme group.)

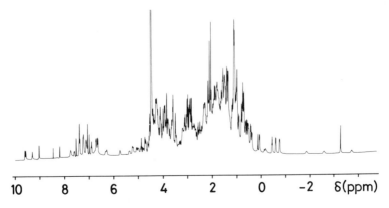

Figure 3.6. ^{1}H NMR spectrum of horse ferrocytochrome c in D_2O (0.006 M, pD 7.2, 53°C; 360 MHz) after complete exchange of the labile protons.

3.2. SYNTHETIC NUCLEIC ACID FRAGMENTS

The recent development of new techniques for rapid and efficient synthesis of polynucleotides with precisely defined sequences enabled X-ray studies in single crystals (Dickerson, 1983) and NMR studies in solution. A wealth of details on the molecular conformations in nucleic acids have thus become accessible for experimental observation. Among the possible conformations formed by synthetic polynucleotides in solution are flexible, "unstructured" single strands, stacked helical single strands, hairpins, regular duplexes formed by complementary strands, and a variety of aggregates between partially complementary strands, which may contain bulges, dangling ends, or stacked single stranded ends. Conformational studies by NMR may include determination of the stoichiometry of the species present under specified solution conditions, identification of the conformation type in these species, and investigations of static and dynamic details of the spatial structures.

Figure 3.7 shows two ^{1}H NMR spectra of the self-complementary hexa-deoxyribonucleotide pentaphosphate d(GCATGC). The spectrum at 55 °C corresponds to a predominantly single stranded, flexible form of the molecule, and at 20°C the predominant species is a regular duplex. In both spectra the distribution of resonance intensity along the chemical shift axis corresponds to the expectations from Table 2.6. Most resonance lines have different chemical shifts in the single strand and in the duplex, and the lines in the

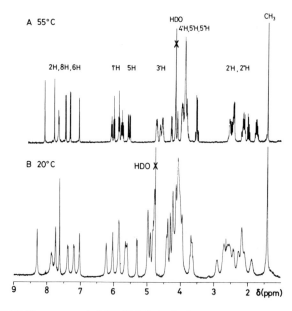

Figure 3.7. ^{1}H NMR spectra of d(GCATGC) in D$_2$O (0.004 M, pD 7.0, 0.05 M phosphate, 0.1 M NaCl; 360 MHz). (A) Single strands. (B) Duplex.

duplex are broadened. Factors contributing to this line broadening include that the duplex structure is comparatively rigid, with limited segmental mobility for the individual strands, and the increased solvent viscosity at the lower temperature. Comparison of d(GCATGC)$_2$ (Fig. 3.7B) and d(GCAT-TAATGC)$_2$ (Fig. 3.8A) affords an illustration of the increase of complexity of the ^{1}H NMR spectrum with increasing molecular size. For the decanucleotide duplex two regions of the spectrum in H$_2$O are also shown (Fig. 3.8B), which contain additional lines corresponding to labile protons. These are again in the chemical shift ranges expected from Table 2.6.

Figures 3.7 and 3.8 show that multiple copies of chemically identical monomers have different chemical shifts in these oligonucleotides. This is particularly clearly exemplified by the three methyl lines of T4, T5, and T8 in d(GCATTAATGC)$_2$ (Fig. 3.8). The imino protons manifest the twofold symmetry of the duplex structure formed by this self-complementary decanucleotide, since only 5 (rather than 10) separate imino proton resonances are observed, whereby each resonance corresponds to two conformationally equivalent hydrogens.

Qualitatively, the chemical shift dispersion in nucleic acid fragments can be rationalized along similar lines as those followed for proteins. For these

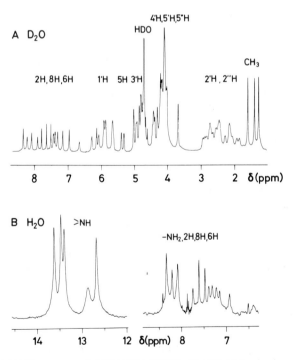

Figure 3.8. ^{1}H NMR spectra of d(GCATTAATGC)$_2$ (0.002 M in duplex, 0.05 M phosphate, 0.1 M NaCl, pH 7.0; 500 MHz). (A) In D$_2$O, 28°C. (B) In H$_2$O, 15°C, solvent suppression by a semiselective observation pulse.

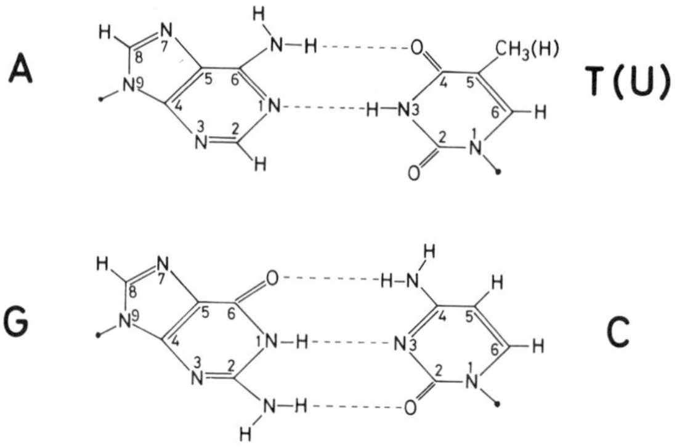

Figure 3.9. The Watson–Crick base pairs A=T (A=U) and G≡C.

relatively short chains, end group effects may also be quite important (e.g., the well-separated resonance at the high field end of the 4'H–5'H region in Fig. 3.7A comes from 5'CH$_2$ of G1, which is not phosphorylated). For non-terminal residues the dominant influences on the shifts come from the ring current fields of the bases and from hydrogen bonding. Because of the hete-roatoms and the polar substituents, the ring current fields of pyrimidines and purines are more complex than that of a benzene ring (Fig. 3.5), but these have been investigated extensively and ring current shift contours are avail-able (e.g., Giessner-Prettre and Pullman, 1970; Giessner-Prettre, 1984). There are two other factors that led to somewhat different methods of analy-sis of conformation-dependent shifts in proteins and in nucleic acids. First, in contrast to proteins, ring current fields and polar groups are omnipresent in nucleotides, so that their influence causes sequence-dependent chemical shifts even in "unstructured," flexible oligonucleotides. Second, the build-ing blocks of regular nucleic acid duplexes are the Watson–Crick base pairs (Fig. 3.9) rather than the individual mononucleotides. *Intrinsic chemical shifts* for the individual *base pairs* were obtained from a combination of NMR measurements and computation of the ring current effects from the neighbor and nearest-neighbor base pairs (e.g., Bell et al., 1985; Hilbers, 1979; Robillard and Reid, 1979). For example, the intrinsic chemical shifts for the imino protons in the Watson–Crick base pairs A=U (or A=T) and G≡C were thus found to be approximately 14.3 and 13.5 ppm, respectively.

3.3. TRANSFER RNA

Historically, systematic NMR studies of tRNA's provided many of the fun-damental ideas that also proved fruitful for the aforementioned studies of nucleic acid fragments. This includes, for example, the observation of hy-drogen-bonded labile base protons (Kearns et al., 1971) and the ring current

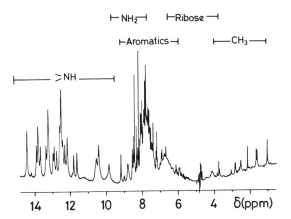

Figure 3.10. ^{1}H NMR spectrum of yeast tRNA^{Phe} in H_2O (0.1 M NaCl, 0.005 M MgCl$_2$, pH 7.0, 28°C; 500 MHz). The solvent resonance was suppressed by the combined application of a semiselective observation pulse and digital shift accumulation (from Hilbers et al., © 1983 by D. Reidel Publishing Company, Dordrecht, Holland).

calculations for the analysis of chemical shifts (Robillard and Reid, 1979). Initially the NMR studies concentrated primarily on the high field region from approximately 0 to 4 ppm, which contains methyl and methylene proton resonances, and the low field region from 10 to 15 ppm, which contains resonances of hydrogen-bonded imino protons (Fig. 3.10). More recently, with the advent of higher magnetic fields and new experimental procedures, the studies were extended to the region containing the amino protons and the nonlabile base protons. In the spectrum of Figure 3.10 the intensity of the ribose resonances is greatly attenuated as an artifact of the solvent suppression. However, even in spectra recorded in D_2O, this region was so far only sparsely exploited because of the complexities arising from overlap of several hundred lines.

When trying to rationalize the ^{1}H NMR spectra of tRNA's, one finds elements previously encountered in nucleic acid fragments and proteins as well as features that are unique for tRNA. Similar to synthetic oligonucleotide duplexes tRNA's contain numerous Watson–Crick type base pairs in the stem regions, which give rise to imino-proton resonances at the low field end of the spectrum and to amino-proton lines between 8 and 10 ppm (Fig. 3.10). Unique spectral features arise from the presence of "rare" nucleotides with modified chemical structures, and from the formation of hydrogen bonds that do not correspond to Watson–Crick base pairs (Fig. 3.11). Of particular interest are nucleotides modified by methylation (e.g., m^{7}G, m^{1}A, and m$^2_2$G in Fig. 3.11), since none of the common RNA nucleotides contain methyl groups (Table 2.5). Unusual hydrogen bonding includes Hoogsteen and reversed Hoogsteen base pairs, formation of triples, and involvement of backbone groups (Fig. 3.11), and may be manifested by

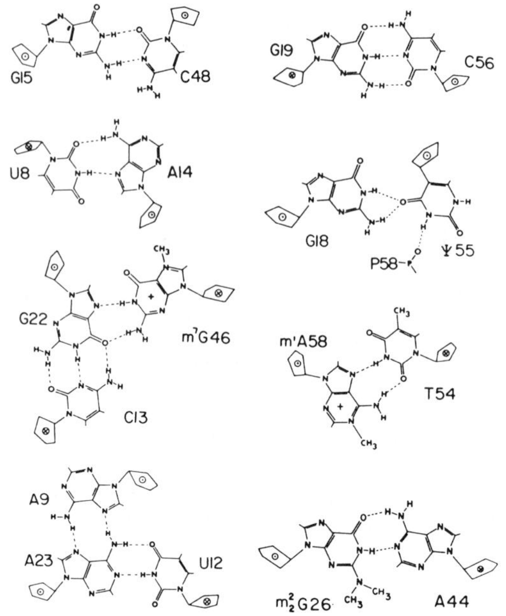

Figure 3.11. Rare nucleotides and tertiary structure hydrogen bonds in yeast tRNA^{Phe}. (Reproduced with permission from Rich and RajBhandary, *Annual Reviews of Biochemistry*, Volume 45, ©1976 by Annual Reviews Inc.)

additional lines in the spectral region from 8 to 15 ppm (Hilbers, 1979; Robillard and Reid, 1979). Similar to proteins the tertiary structure of tRNA's is manifested by upfield ring current shifts of methyl resonances. A nice illustration is afforded by comparing the methyl regions in the ¹H NMR spectra of native and heat-denatured yeast tRNA^{Phe} (Fig. 3.12). The well separated methyl resonances in folded tRNA provide a naturally built-in set of easily observable NMR probes in these relatively large molecules.

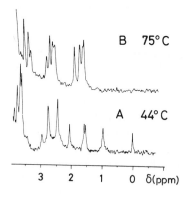

Figure 3.12. Chemical shift dispersion for methyl ¹H NMR lines in yeast tRNA^{Phe} (D$_2$O, 0.1 M NaCl, 0.015 M MgCl$_2$, pD 7.0; 360 MHz). *(A)* Native tRNA^{Phe} *(B)* Denatured tRNA^{Phe} (adapted from Robillard and Reid, 1979).

3.4. NMR STUDIES OF LABILE PROTONS

In Chapter 2 we identified those protons in amino acid residues and nucleotides that are *labile* in the sense that they can readily be exchanged with solvent protons in aqueous media (Tables 2.2 and 2.5). So far, in Chapter 3, we evaluated the use of NMR observation of labile protons in proteins and nucleic acids. In the following we further discuss some practical aspects of NMR spectroscopy with labile protons.

Because of the presence of labile protons, the ^1H NMR spectra of proteins and nucleic acids are simplified in two ways by changing the solvent from H_2O to D_2O. First, the H_2O solvent resonance (110 M in protons) near 4.8 ppm is replaced by the much weaker line of the residual protons in D_2O (typically 0.1 M in protons). Second, the resonance lines of the labile protons disappear, and so do the spin–spin couplings with labile protons in the remaining resonance lines of nonlabile protons.

For nucleic acids complete exchange is usually achieved by dissolving the compound in D_2O. For proteins, complete exchange is obtained under nondenaturing conditions by keeping a D_2O solution at elevated temperature for several minutes or hours. After the exchange the molar solvent proton concentration is two to three orders of magnitude higher than that of the biopolymer, so that proper sample preparation requires further repeated replacement of the solvent by D_2O, for example, by lyophilization or by dialysis.

In a freshly prepared protein solution in D_2O some of the labile protons can often be observed. For example, for BPTI at 36°C and pD 4.6, 33 labile protons were thus observed (Wüthrich and Wagner, 1979). These are all located in the interior of the protein, whereas surface protons exchange too rapidly to be seen in D_2O. (The limiting value of the exchange rate constant k_m, which enables observation in D_2O, is of the order 0.1 min^{-1}. This allows for the time needed for sample preparation, instrument setup, and recording of the spectrum.) By observation of the interior labile protons at variable times after the sample preparation, information on the exchange kinetics can be obtained (Fig. 3.13).

For surface amide protons that are freely accessible to the solvent, the exchange rates are comparable to k_{intr}. In the pH range 1–6 they are thus slower than approximately 10 min^{-1} (Fig. 2.3). While this is too fast for observation in D_2O, it is slow on the NMR time scale defined by the chemical shifts relative to H_2O, so that all amide protons can be observed in H_2O solutions. At pH below approximately 6.0 the strong water resonance can in such experiments be suppressed by selective saturation.

In analogy to simplified subspectra containing only slowly exchanging protons (Fig. 3.13), subspectra containing only the rapidly exchanging surface protons may be obtained in H_2O solutions of proteins for which all labile protons had previously been replaced by deuterium (Keller et al., 1983; Tüchsen and Woodward, 1985a).

In Section 2.3 we estimated that the NMR time scale determined by the

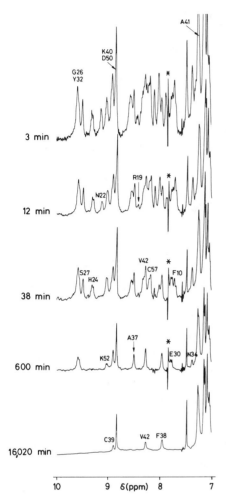

Figure 3.13. Observation of amide-proton exchange in the ^{1}H NMR spectra of the small protein bull seminal inhibitor (BUSI) in D$_2$O (0.006 M, pD 4.1, 25°C; 360 MHz). The individual spectra were recorded at the indicated times after sample preparation. All resonance assignments indicate backbone amide-proton lines, * identifies an instrumental artifact (from Štrop and Wüthrich, 1983).

relative chemical shifts of labile protons in proteins or nucleic acids and the solvent water allows observation of protons exchanging with rates up to approximately 1×10^3 min^{-1}. In the range $k_m \approx 10-1 \times 10^4$ min^{-1} we have the intrinsic exchange rates for amide protons in the pH range 6–9 (Fig. 2.3) and the exchange rates for hydrogen-bonded labile protons in nucleic acids (Hilbers, 1979). Thus, surface amide protons in protein solutions at basic pH and the hydrogen bonds in nucleic acids can also be observed in H$_2$O solution. However, in this range of exchange rates care must be taken to prevent loss of resonance intensity due to *saturation transfer*, since the longitudinal relaxation times T_1 of the labile protons in macromolecules may be long compared to their lifetimes with respect to exchange with the solvent. If in this situation the water resonance is suppressed by saturation, the labile

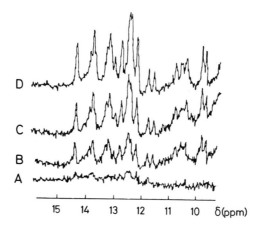

Figure 3.14. Transfer of saturation between solvent H_2O and the labile protons of yeast tRNA^{Phe} in ^1H NMR spectra in H_2O (0.1 M NaCl, 0.001 M MgCl$_2$, 0.007 M EDTA, 0.01 M sodium cacodylate, pH 7.0, 35°C; 270 MHz). (A) Saturation of the H_2O resonance. (B) and (C) Partial saturation of H_2O by off-resonance irradiation. (D) No saturation (from Johnston and Redfield, 1977).

protons in the macromolecules will be replaced by saturated spins from the huge spin reservoir of the solvent, with concomitant loss of resonance intensity. In the example of Figure 3.14A, the resonances of all labile protons in yeast tRNA^{Phe} were nearly completely eliminated by saturation transfer from the solvent. Such saturation transfer experiments can be employed for quantitative investigations of the proton exchange (Hoffman and Forsén, 1966; Johnston and Redfield, 1977). However, for high resolution NMR studies with labile protons exchanging with rates between 10 and 1×10^4 min^{-1}, different solvent suppression schemes must be applied. These rely primarily on the use of a *semiselective observation pulse* (see Fig. 5.2 and Section 5.7).

CHAPTER 4
The NMR Assignment Problem in Biopolymers

It has long been recognized that the information gained from NMR measurements is largely determined by the extent to which the resonance lines can be assigned (Section 1.2), and much space in the literature was devoted to this problem. Historically, inability to resolve individual lines in the NMR spectra of biological macromolecules was a major stumbling block up to the early 1970s, but ever since the lack of generally practicable, efficient methods for assignment of resolved lines has been a prime limiting factor. On the basis of the material on structure and NMR spectra of proteins and nucleic acids presented in Chapters 2 and 3 we are now in a position to further evaluate fundamental and practical aspects of NMR assignments.

Let us first consider the nuclear spin systems of polypeptide and polynucleotide chains. Thereby, we must allow for the practical limitation that observation of $^1H-^1H$ couplings in biopolymers is with few exceptions limited to protons separated by two covalent bonds (*geminal* couplings, ^2J) or three covalent bonds (*vicinal* couplings, ^3J). Inspection of polypeptide and polynucleotide covalent structures then shows that these contain the same ^1H spin systems as the constituent monomeric building blocks (Tables 2.2 and 2.5). Experiments capable of delineating scalar $^1H-^1H$ couplings will therefore only enable classification of the resonance lines in groups corresponding to ^1H spin systems in individual monomeric units (sequence-specific assignments may result when the polymer contains single copies of unique monomers, for example, an unusual nucleotide in tRNA). In contrast to the homonuclear ^1H spin systems, heteronuclear systems with the isotopes of Table 2.1 can be constructed that extend over the entire biopolymer chain. Thereby, as was already mentioned in Chapter 2, the systems containing ^{15}N or ^{13}C at natural abundance will be present only in very low concentrations. However, with regard to potential future developments of NMR techniques it is important to keep in mind that any unique biopolymer sequence corresponds to a unique heteronuclear spin system, so that complete sequence-specific resonance assignments could in principle be obtained entirely from observation of scalar spin–spin couplings.

Since the NMR spectral resolution of multiple copies of identical monomeric units in proteins or nucleic acids relies on the dispersion of the chemical shifts by the spatial folding of the polymer chains (Figs. 3.3 and 3.12), application of certain assignment strategies that are widely used in organic chemistry is limited on fundamental grounds. This includes the use of empirical rules relating chemical shifts and spin–spin coupling constants with chemical structures (e.g., Emsley et al., 1966; Stothers, 1972), chemical modification of the molecular structure, and incremental approaches relying on comparison of the intact molecule with small chain fragments obtained by chemical or biochemical methods (e.g., Borer et al., 1975). The important conclusion is that *the resonance assignments must be obtained from NMR spectra of the intact macromolecules recorded with nondenaturing conditions.*

Two strategies capable of providing sequence-specific resonance assignments under these naturally given conditions, namely, sequential assignments by NMR techniques and isotope labeling are outlined in Sections 4.1 and 4.2. Many different assignment procedures were tried out in the past (for reviews see, e.g., Jardetzky and Roberts, 1981; Markley and Ulrich, 1984; Patel et al., 1982a; Robillard and Reid, 1979; Wüthrich, 1976). These include comparison of observed chemical shifts with shifts obtained from ring current calculations with the corresponding crystal structure, comparative studies with homologous molecules or chemically modified analogs, assignments with NOE's based on internuclear distances derived from the crystal structure, use of paramagnetic shift or relaxation reagents in conjunction with the crystal structure, or assignments of labile protons on the basis of the kinetics of exchange with the solvent. A limitation common to all these procedures is that they can usually provide assignments only for a small fraction of the molecular structure (e.g., the peripheral side chain hydrogens in proteins or the methyl lines in tRNA's). Furthermore, they rely on information (or assumptions) about conformation, dynamics, or intermolecular interactions in the molecules of interest, which leads almost inevitably to circular arguments in further interpretations of the NMR data. Nonetheless, it is very important that these procedures were tried out. At the time they were in most cases the optimum that was practicable with the available equipment, and usually they added new insights into NMR spectral features of biopolymers. In special cases they also provided reliable resonance assignments. [An example is found in the hemoproteins, where early assignments based on unusual chemical shifts of protons located near the heme groups provided a basis for extensive investigations on structure–function correlations; for reviews see, e.g., La Mar (1979), Wüthrich (1976).] However, for use with biopolymers in general the available literature unfortunately documents clearly that these "less rigorous methods" (quotation from Jardetzky and Roberts, 1981, p. 256) did often not provide the desired assignments.

4.1. SEQUENTIAL ASSIGNMENTS

Sequential assignments make use of the fact that unique oligopeptide or oligonucleotide segments, respectively, are ubiquitous in the primary structures of proteins and nucleic acids (in proteins, unique chain segments typically comprise two to four residues; see Chapter 8). In a sequential assignment procedure, connectivities between the ^{1}H spin systems of sequentially neighboring monomeric units are established either by heteronuclear scalar couplings or by ^{1}H–^{1}H NOE's. The resulting short segments of neighboring ^{1}H spin systems are then matched against the corresponding segments in the chemically determined sequence of the biopolymer chain to obtain sequence-specific assignments. Sequential assignments for proteins and nucleic acids are described in detail in Chapters 8, 12, and 13.

4.2. ISOTOPE LABELING

Through isotope labeling the NMR spectra can be modified without noticeably affecting molecular structure and function. In principle this enables a straightforward solution of the assignment problem in biopolymers: Isotope labeling of a particular monomeric unit affects its spin system, so that it can be uniquely recognized. As an illustration, consider the impact of ^{15}N enrichment at the amide nitrogen position of an amino acid residue in a polypeptide chain. First, in the ^{15}N NMR spectrum the resonance of the labeled residue could be assigned from its outstanding intensity. Second, the ^{1}H spin system of the labeled residue could be identified from the doublet fine structure of the amide-proton resonance that arises from scalar ^{1}H–^{15}N coupling. Third, in the ^{13}C NMR spectrum the α-carbon resonance of the labeled residue and the backbone carbonyl-carbon line of the preceding residue could be identified from the scalar ^{15}N–^{13}C couplings.

Clearly, an ideal assignment procedure for biopolymers could be based on the production of suitably enriched monomers followed by incorporation of individual labeled monomers into specified sequence positions. Both steps are difficult and laborious, but progress is being made. For example, methods for preparative scale production of amino acids with site-specific isotope enrichment have been described (Le Master and Richards, 1982) and Lys with ^{13}C enrichment at the carbonyl-carbon position was introduced by enzymatic techniques into the reactive site in soybean trypsin inhibitor and in

BPTI for studies of the protein–protein interactions in the protease–inhibitor complexes (Baillargeon et al., 1980; Hunkapiller, et al. 1979; Richarz et al., 1980b). For more extensive resonance assignments, double labeling on the level of amino acid types appears more promising than introduction of single labeled residues into specified sequence locations. For example, in a protein containing Ala labeled at the carbonyl carbon and Gly labeled at the amide nitrogen, all segments Ala-Gly could be identified from the ^{13}C–^{15}N scalar couplings. Combining this result with those from different double labeling experiments (e.g., identifying the dipeptide segments Leu-Ala and Gly-Val) should eventually result in sequence-specific resonance assignments (Kainosho and Tsuji, 1982).

Isotope substitution is widely used for simplifying NMR spectra or facilitating observation of ^{13}C or ^{15}N by nonselective enrichment. While this does not in itself provide resonance assignments, it may provide important support for assignment procedures (see also Section 3.4). In a pioneering project, Markley et al. (1968) used biosynthetic methods for producing proteins containing all but one amino acid type (or a selected few) in perdeuterated form. In such simplified ^1H NMR spectra the spin systems of the individual residues can be more easily identified using the same NMR techniques as for the fully protonated molecule. Following a different strategy, it was also proposed to perdeuterate biopolymers to the extent of approximately 90% and to observe the complete ^1H NMR spectrum of the residual 10% protons (Kalbitzer et al., 1985). As a principal advantage of this procedure the line broadening through ^1H–^1H dipole–dipole coupling (Fig. 3.1) would be reduced by the isotope dilution of the protons in the macromolecular structure, and the intrinsic loss in sensitivity would be at least partially offset by the narrower lines.

Overall, extensive NMR assignments by selective isotope labeling appear too laborious to be widely practicable. However, it can be foreseen that the range of practical applications for sequential assignment procedures could be enhanced by a combination with nonselective labeling, for example, by labeling of amino acid types in proteins (Chapters 8, 12, and 13).

CHAPTER 5
Two-Dimensional NMR with Proteins and Nucleic Acids

At the outset of this chapter, a brief look at one-dimensional (1D) NMR will serve both as a general introduction and as a reference for appreciating the advantages of two-dimensional (2D) NMR for studies of biological macromolecules. Figure 5.1A presents an experimental scheme for 1D Fourier transform (FT)-NMR. After an equilibrium magnetization **M** parallel to $\mathbf{B}_0$ has been established by placing the sample in the static magnetic field, a radio-frequency (rf) pulse $\mathbf{B}_1$ is applied that rotates **M** away from the z axis (Fig. 1.4A) by a *flip angle* β. Immediately after the pulse the *free induction decay* (FID) is recorded. In a standard experiment a *nonselective* 90° pulse is applied, by which the magnetization resulting from the ensemble of all spins of a given type (e.g., all protons) is rotated into the x,y plane (Fig. 1.4). In typical experiments with biopolymers, the same measurement is repeated n times and the FID's are coadded to improve the *signal-to-noise ratio (S/N)*. 1D FT of the FID then yields the complete 1D NMR spectrum in the frequency domain (e.g., Figs. 3.3B and 3.6–3.8).

The introductory remarks on NMR presented in Section 1.4 lead to some important conclusions relating to the experiment in Figure 5.1A, which are also relevant for 2D NMR.

1. The envelope of the FID (Fig. 5.1A) is determined by the inverse of the apparent transverse relaxation time, $1/T_2^* = 1/T_2 + 1/T_2^+$.

2. The repetition rate is limited by the longitudinal relaxation time T_1, since the equilibrium magnetization along the z axis must be reestablished before the next measurement.

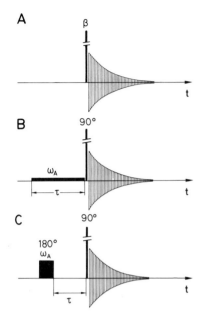

Figure 5.1. Experimental scheme of 1D FT-NMR experiments. β or 90° indicate a nonselective observation pulse with flip angle β or 90°, respectively. t indicates the time axis along which the FID is observed. *(A)* Normal 1D FT-NMR. *(B)* Truncated driven NOE (TOE); selective saturation of resonance A by irradiation during the period τ immediately before the observation pulse. *(C)* Transient NOE; selective 180° pulse on resonance A at a time τ before the observation pulse.

3. The flip angle of the magnetization is proportional to the amplitude of the rf field $\mathbf{B}_1$ and the duration of the pulse. At constant amplitude of $\mathbf{B}_1$, the pulse length can be varied so as to produce, for example, a 45, 90, or 180°-pulse.

4. The *phase* of the rf pulse is determined by the orientation of $\mathbf{B}_1$ relative to the rotating coordinate system (Fig. 1.4A). Pulses may, for example, be applied along x', y', $-x'$, or $-y'$, which means that they are *phase shifted* by 0, 90, 180, or 270° relative to the receiver reference phase.

5. The acquisition of the FID may start immediately following the observation pulse *(direct acquisition)* or after a specified waiting time *(delayed acquisition)*.

6. The nonselective observation pulse can be replaced by *selective excitation* schemes. For example, the symmetric gated excitation sequences of total length τ with coherent radio frequency f_1 shown in Figure 5.2 all have nulls in their power spectra at frequencies $f_1 \pm \tau^{-1}$. By proper adjustment of the difference between the resonance frequency of the solvent and f_1, a spectrum can be recorded with strong attenuation (approximately 1:100) of the solvent signal (and the spectral region near the solvent; Fig. 3.10). A variety of related selective excitation schemes are available (e.g., Clore et al., 1983; Hore, 1983).

A normal 1D NMR experiment (Fig. 5.1A) provides information on the chemical shifts and the spin–spin coupling fine structures of the individual

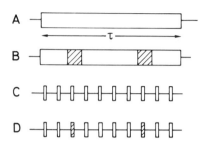

Figure 5.2. Long-pulse excitation schemes used for solvent suppression in H_2O solutions. The total sequence length τ is of the order 1 ms and varies depending on the spectral region studied. Hatched sections are 180° out of phase relative to the rest of the sequence. (A) Long pulse. (B) 214 pulse. (C) Time-shared long pulse. (D) Time-shared 214 pulse. [Reprinted from Redfield and Kunz in *NMR and Biochemistry*, p. 230, by Courtesy of Marcel Dekker, Inc. N.Y. (1979)].

resonances in the spectrum. To obtain additional data on through-bond, scalar connectivities or through-space, dipolar connectivities between individual spins, which provide the basis for resonance assignments and conformational studies in biopolymers, double or multiple irradiation experiments must be used. These rely on *selective* irradiation of a particular resonance line with a rf field $\mathbf{B}_2$ and observation of the resulting effects in the rest of the spectrum. For nuclear Overhauser enhancement (NOE) experiments the selective irradiation is applied prior to the observation pulse (Fig. 5.1B and C), and for spin decoupling experiments it is applied in a time-shared mode (Fig. 5.2C and D) during the acquisition of the FID. For work with the complex, crowded spectra of biopolymers (Figs. 3.6–3.8 and 3.10) the use of 1D double irradiation experiments is naturally limited. While the spectral resolution for observation of double irradiation effects in crowded spectral regions has customarily been improved by difference spectroscopy (De Marco et al., 1977; Gibbons et al., 1975; Johnston and Redfield, 1978; Richarz and Wüthrich, 1978b), lack of selectivity for irradiation of individual lines in spectral regions with mutually overlapping resonances is a stringent limitation. Furthermore, since each 1D double resonance experiment provides only one connectivity (in favorable cases perhaps two to five), a very large number of measurements would be required for characterization of the complete network of spin–spin connectivities in a macromolecular structure.

With 2D NMR techniques these natural limitations of 1D NMR can be largely overcome. A 2D NMR experiment is recorded in a 2D time space in the following sense. Analogous to 1D NMR (Fig. 5.1A) the FID is recorded during the *detection period* t_2 after the observation pulse. However, taking homonuclear 2D correlated spectroscopy (COSY) as an example (Fig. 5.3A), another nonselective rf pulse is applied prior to the observation pulse. The influence of this first pulse on the FID recorded during t_2 depends on the length of the *evolution period* t_1 between the two pulses. A second time dimension can thus be created by repeating the same experiment with incrementation of t_1. For each value of t_1 a FID recorded versus t_2 is stored, so that a data matrix $s(t_1,t_2)$ is obtained. A 2D FT of $s(t_1,t_2)$ then produces the

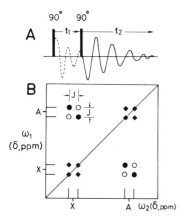

Figure 5.3. *(A)* Experimental scheme for homonuclear COSY. 90° indicates nonselective 90° pulses. t_1 is the *evolution period*. t_2 is the *observation period* during which the FID is recorded. *(B)* Schematic, phase-sensitive presentation of a COSY spectrum for an AX spin system. Positive absorptive, negative absorptive, and dispersive peaks are indicated by filled circles, open circles, and diamonds, respectively. Stick diagrams of the corresponding 1D NMR spectrum are indicated along the ω_1 and ω_2 axes.

desired 2D frequency spectrum $S(\omega_1,\omega_2)$ (Fig. 5.3*B*). In the COSY spectrum of Figure 5.3*B* the chemical shift information is contained in the positions of the two doublets on the diagonal from the upper right to the lower left *(diagonal peaks)*. Scalar coupling connectivities between individual spins are manifested by *cross peaks* located at the intersections of straight lines parallel to the frequency axes ω_1 and ω_2 through the diagonal peaks. The spin–spin coupling fine structure is shown for the diagonal peaks and the cross peaks. For work with biological macromolecules, the most important, fundamental advantage of 2D NMR relative to 1D NMR is that experiments using exclusively nonselective rf pulses (Fig. 5.3*A*) can delineate connectivities between distinct individual spins. In addition, a single COSY experiment can in principle delineate all scalar spin–spin coupling connectivities between protons in a macromolecular structure, and it is thus much more efficient than the use of 1D spin-decoupling experiments. Furthermore, because the resonance peaks are spread out in two dimensions, the peak separation is usually substantially improved.

In the following Sections 5.1–5.8 and in Section 6.3 the emphasis is on a phenomenological description of appearance, analysis, and information content of 2D NMR spectra of proteins and nucleic acids. More comprehensive presentations of 2D NMR spectroscopy can be found in the books by Bax (1982) and by Ernst et al., (1987), as well as in a variety of review papers (e.g., Freeman and Morris, 1979; Nagayama, 1981; Sørensen et al., 1983; Wider et al., 1984). It is most important, however, to keep in mind that both NMR instrumentation and experimental techniques are rapidly developing and that the material presented here and in the preceding references should also serve as a basis for access to the current original literature. In this sense the material in this chapter was selected for the dual purposes of awakening an understanding for the selection of the experiments used in the past, and creating a sense of anticipation for further improvements in the future.

5.1. EXPERIMENTAL SCHEMES FOR 2D NMR SPECTROSCOPY

A general scheme for 2D NMR includes four successive time periods, *prepa-ration, evolution, mixing,* and *detection* (Fig. 5.4*A*). Preparation, evolution, and detection are mandatory, and most 2D NMR experiments include a *mixing period* or a *mixing pulse*. The preparation period usually consists of a *delay time*, during which thermal equilibrium is attained (Fig. 1.4*A*), fol-lowed by one or several rf pulses to create *coherence*. [Coherence can mean phase coherence among like or unlike spins. It includes the concept of transverse magnetization (Fig. 1.4*B*), corresponding to observable single-quantum transitions, where the majority of spins in an ensemble of like spins have the same phase in the $x'y'$ plane. In the context of, not directly observ-able, zero-quantum, 2-quantum, or higher-order coherence, it means that in the majority of molecules two or several spins of different types have the same phases relative to each other.] During the evolution period the coher-ence evolves, and at the end of this interval the system assumes a specified state (Fig. 5.3*A*), which depends on the elapsed time t_1 and the Hamiltonian operator $\mathcal{H}^1$ effective during t_1. The mixing period may include one or sev-

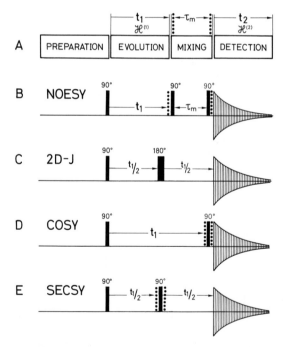

Figure 5.4. *(A)* General experimental scheme for 2D NMR. *(B)–(E)* Experimental schemes for four homonuclear ^{1}H 2D NMR experiments. The dotted lines indicate the bounds of the mixing period (or mixing pulse) in the different experiments. 2D-*J* spec-troscopy has no mixing period.

eral rf pulses and delay intervals. During the mixing period, coherence is transferred between spins [e.g., in autocorrelation experiments, the mixing process determines the frequency pairs (ω_1, ω_2) with cross peaks of nonvanishing intensity]. During the detection period the system evolves further (Fig. 5.3A) under the Hamiltonian operator $\mathcal{H}^2$ effective during t_2, and the resulting FID is recorded and stored as $s^*(t_2)$. The lengths of the preparation, mixing, and detection periods are in principle invariant for all the measurements $s^*(t_2)$, whereas t_1 is incremented between the individual measurements $s^*(t_2)$ to obtain a complete 2D data set $s(t_1, t_2)$. In the 2D spectrum $S(\omega_1, \omega_2)$ obtained by 2D FT of $s(t_1, t_2)$ (Fig. 5.3B), the precession frequencies during evolution and detection (i.e., under $\mathcal{H}^1$ and $\mathcal{H}^2$, respectively) determine the coordinates of the signal peaks.

2D NMR has a wide range of applications for homonuclear and heteronuclear experiments in solution, liquid crystals, and solids. The general scheme of Figure 5.4A applies to all potential applications. By suitable combination of different preparation, evolution, mixing, and detection schemes, a wide spectrum of 2D experiments can be designed, which may be taylored to particular requirements dictated by the problem to be solved. Figure 5.4B–E presents a small selection of homonuclear 2D NMR experiments, which have been of central importance for the early development of ^{1}H 2D NMR studies with biopolymers. A variety of additional 2D experiments will be used throughout the text, which can all be generated from the scheme of Figure 5.4A.

The four experiments presented in Figure 5.4B–E differ in evolution and mixing. In all cases the preparation consists of a delay interval, which can be adjusted according to T_1 of the spins of interest, and a 90° pulse at the end. The detection is the same in all four pulse schemes. In 2D nuclear Overhauser enhancement and exchange spectroscopy (NOESY) the mixing period consists of two 90° pulses separated by the *mixing time* τ_m (Fig. 5.4B). The cross peaks in NOESY correlation maps put into evidence dipole–dipole coupling between nuclear spins in close spatial proximity, which gives rise to NOE's (Anil-Kumar et al., 1980a). The same experiment can be employed for studies of magnetization transfer by chemical exchange, or by physical exchange processes such as the flipping motions of aromatic rings in proteins (Figs. 1.5 and 1.6) (Jeener et al., 1979; Meier and Ernst, 1979; see Chapter 6).

In two-dimensional J-resolved spectroscopy (2D-J spectroscopy) the evolution period contains a 180° pulse in the middle, and there is no mixing period (Fig. 5.4C). At the end of the evolution a *spin echo* is produced, which for weakly coupled spin systems manifests exclusively the spin–spin couplings (Hahn and Maxwell, 1952). 2D-J spectroscopy thus affords a separation of spin–spin coupling fine structure and chemical shifts, so that after suitable data manipulation the resonance multiplets are manifested along the ω_1 axis perpendicular to the chemical shift axis ω_2 (Aue et al., 1976b; Nagayama et al., 1977, 1978). Figure 5.5 illustrates with the region of the methyl

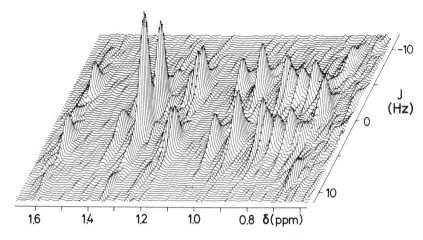

Figure 5.5. ^{1}H 2D-J spectrum of BPTI (0.01 M, D_2O, pD 4.5, 60°C; 360 MHz). A stacked plot of the high field region is shown, containing 17 methyl doublets and 2 overlapped methyl triplets.

doublet and triplet resonances in BPTI that the spectral resolution can be greatly improved by the 2D-J technique, and that this experiment readily identifies all components of a given multiplet.

In COSY the mixing is obtained by the 90° pulse separating evolution and detection (Fig. 5.4D). As was already illustrated with Figure 5.3B, COSY produces correlation maps that display connectivities by scalar spin–spin coupling and thus provide information on proximity of nuclei along the chemical bonds (Aue et al., 1976a; Bax and Freeman, 1981; Nagayama et al., 1980).

2D spin-echo correlated spectroscopy (SECSY) (Fig. 5.4E) is a variant of COSY using *delayed detection* (Nagayama et al., 1979; 1980). Formally it resembles the 2D-J experiment, from which it can be derived by replacing the 180° spin-echo pulse by a 90° mixing pulse. Delayed acquisition results in a different outlay of the correlation map (Fig. 5.6). The peaks in a SECSY

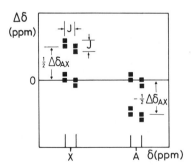

Figure 5.6. SECSY spectrum for an AX spin system, absolute-value presentation.

spectrum which correspond to the diagonal peaks in COSY (Fig. 5.3*B*) are on a horizontal line through the center of the spectrum, and the pairs of cross peaks in symmetrical positions relative to the diagonal in COSY (Fig. 5.3*B*) lie on straight lines at an angle of 45° relative to the horizontal axis (Fig. 5.6). Compared to COSY, a SECSY data set occupies less memory space along ω_1. Another important difference between the two experiments arises because in spectra recorded with delayed acquisition, the absorption and dispersion mode components cannot routinely be separated. It may be added that other 2D NMR experiments besides COSY can be recorded with delayed acquisition, or that 2D-*J* spectra can be obtained with the detection following immediately after the 180° pulse. In all instances these modifications affect the memory requirements along ω_1, which can also be reduced with the use of 2D fold-over corrected spectroscopy (FOCSY) (Nagayama et al., 1980; Wider et al., 1984).

Individual authors have proposed different classifications of 2D NMR experiments according to the underlying principles of spectroscopy (Bax, 1982; Ernst, 1982; Ernst et al., 1987; Freeman and Morris, 1979). For the applications in this book it is most useful to introduce the following classification:

1. Experiments for delineating *through-bond,* scalar spin–spin connectivities. These include COSY, SECSY, and additional homonuclear and heteronuclear experiments introduced in Sections 5.5 and 5.8.

2. Experiments for delineating *through-space,* dipolar spin–spin connectivities. These include NOESY and additional experiments presented in Chapter 6.

2D-*J* spectroscopy is not included in either of these two classes. Historically, it was the first 2D experiment used for studies of biopolymers (Nagayama et al., 1977), and early on it was applied to detailed studies of multiplet fine structures and quantitative measurements of spin–spin coupling constants *J* in protein ¹H NMR spectra (Nagayama and Wüthrich, 1981a,b). However, more recent experience has shown that analysis of the COSY cross peak fine structure (Fig. 5.3*B*) in biopolymers is a more efficient way for obtaining the information previously acquired with 2D-*J* spectroscopy.

5.2. GRAPHICAL PRESENTATION OF 2D NMR SPECTRA

In view of the huge information content of 2D NMR spectra, the selection of techniques for visualization of the data is of considerable practical importance. The following presentations use line drawings exclusively and are suitable for documentation in print. For practical work in the laboratory and for slides, colored plots are often very helpful. In computer graphics displays, colored dark–light intensity plots can also routinely be obtained.

For illustration we use the ^{1}H 2D-J spectrum of the amino acid serine in D$_2$O. The fragment αCH–βCH$_2$ contains a strongly coupled ABC ^{1}H spin system (Table 2.2). A three-dimensional view of the spectrum is afforded by a *stacked plot* (Fig. 5.7C), where the third dimension indicates the peak intensities. A survey of the entire spectrum can also be obtained with a *contour plot* (Fig. 5.7D). Through *projection* of the 2D spectrum parallel to one frequency axis, the positions of resonance peaks along the other frequency axis are obtained in a one-dimensional plot (Fig. 5.7B). [For homonuclear 2D-J spectra the projection along the J axis corresponds to a homonuclear broad-band decoupled 1D spectrum, where the projection peaks represent the chemical shifts of the multiplets in the normal 1D NMR spec-

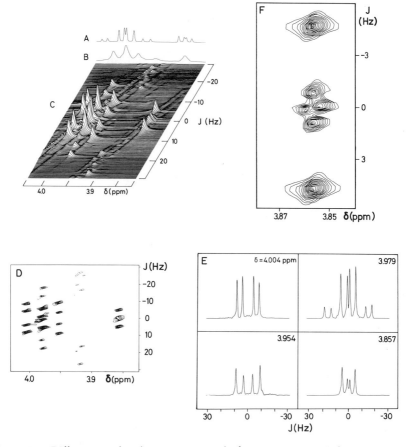

Figure 5.7. Different graphical presentations of a ^{1}H 2D-J spectrum of L-serine (0.1 *M*, D$_2$O, pD 6.5, 25°C; 360 MHz; absolute value). *(A)* 1D NMR. *(B)* Projection of the 2D-J spectrum along the J axis. *(C)* Stacked plot presentation. *(D)* Contour plot. *(E)* Cross sections parallel to the J axis at the chemical shifts of the four most intense projection peaks. *(F)* Enlarged contour plot of the high field multiplet (from Wider et al., 1981).

trum (Fig. 5.7*A*). In Figure 5.7 this is most clearly seen for the highest-field multiplet. The broad lines in the projection are a consequence of the absolute value presentation.] *Cross sections* taken at distinct locations on one frequency axis and along a specified direction afford detailed views of particular spectral features. For example, in Figure 5.7*E*, cross sections through the peaks in the projection spectrum and parallel to the *J* axis show the multiplet fine structures. For detailed views of spectral features that do not lie on straight lines, such as the higher-order effects in the 2D-*J* spectrum of serine (Wider et al., 1981) or the fine structure of COSY cross peaks (Fig. 5.3*B*), contour plots on an expanded scale are most appropriate (Fig. 5.7*F*).

A comparison of stacked plot and contour plot (Fig. 5.7*C* and *D*) illustrates that different criteria must be applied when evaluating a spectrum from different presentations. A stacked plot is a "truth diagram" in the sense that all information is presented, whereas in a contour plot the information below the lowest contour gets lost. For example, in Figure 5.7*D* and *F*, the lowest contour was chosen above the noise level. Generally, a good looking contour plot can be produced from a relatively poor data set by careful selection of the contour levels, and much care must be exercised when extracting quantitative information on peak intensities from contour plots. On the other hand, contour plots are much more suitable than stacked plots for extracting resonance frequencies and delineating connectivities via cross peaks (Fig. 5.3*B*). In practice the spectral analysis relies therefore primarily on contour plots, which are supplemented by cross sections (Fig. 5.7*E*) for proper evaluation of peak heights and *S/N*.

5.3. ¹H COSY SPECTRA OF PROTEINS AND DNA FRAGMENTS

COSY is the standard experiment for studies of scalar spin-spin coupling connectivities. As a first example we consider a spectrum recorded in D_2O solution of the protease inhibitor *K,* which is a homolog of BPTI consisting of 57 amino acid residues. In the stacked plot of the ¹H COSY spectrum (Fig. 5.8*B*), the complete 1D NMR spectrum (Fig. 5.8*A*) can be recognized on the diagonal from the upper right to the lower left. For example, the highest field methyl resonance at -0.9 ppm has the coordinates ($\omega_1 = -0.9$ ppm, $\omega_2 = -0.9$ ppm), and the lowest field amide-proton line at 10.3 ppm is at ($\omega_1 = 10.3$ ppm, $\omega_2 = 10.3$ ppm). The arrangement of the cross peaks is best seen in the contour plot (Fig. 5.8*C*). Except for possible differences between the experimental parameters used for the two frequency dimensions (see Sections 5.4 and 5.5), the spectrum is *symmetrical* with respect to the diagonal. Coupled diagonal peaks can be identified by lines through the cross peaks and parallel to the two frequency axes ω_1 and ω_2. In Figure 5.8*C* this is indicated above the diagonal with broken lines for the connectivities of the highest field Leu methyl resonance with its γH, and the lowest field NH with its αH. The dotted lines show the connectivities for the NH–αH–βH portion

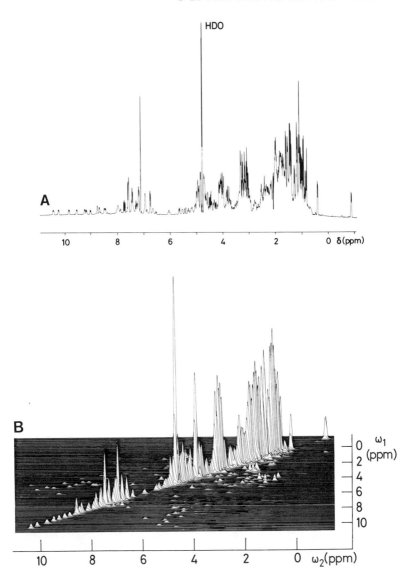

Figure 5.8. ^{1}H COSY of a freshly prepared D_2O solution of inhibitor K (0.01 M, pD 3.4, 25°C; 360 MHz; absolute value). *(A)* 1D ^{1}H NMR. *(B)* Stacked plot of COSY spectrum. *(C)* Contour plot of COSY spectrum. Above the diagonal, broken and dotted lines indicate how connectivities between distinct diagonal peaks are established via the cross peaks. Below the diagonal, solid lines define the regions a–h, which contain cross peaks between distinct proton types (see text). The arrows are explained in the text.

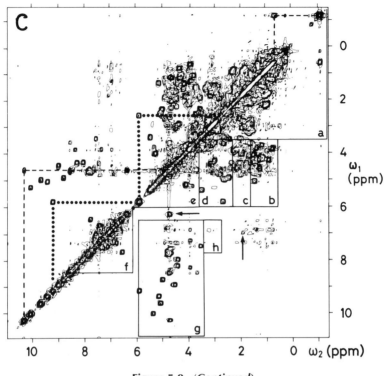

Figure 5.8. (*Continued*)

of a spin system. All cross peaks manifesting connectivities with a particular diagonal peak lie on the cross sections parallel to the ω_1 and ω_2 axes through this diagonal peak.

The COSY spectra displayed in this section are in the absolute value mode and were recorded with low digital resolution of between 5 and 10 Hz/ point. Therefore the fine structure of the resonance peaks (Fig. 5.3*B*) is not resolved. All COSY spectra of biopolymers in the literature up to 1983 were obtained with similar experimental conditions. As a second example, Figure 5.9*A* presents a ¹H COSY spectrum of BPTI recorded in H_2O solution. Comparison with Figure 5.8*B* shows that besides some differences arising from the different amino acid compositions in BPTI and inhibitor *K*, the BPTI spectrum has narrower lines because it was recorded at a higher temperature. Furthermore, it contains a larger number of peaks corresponding to labile protons, and near ($\omega_1 = 5$ ppm, $\omega_2 = 5$ ppm) there are residual perturbations from the water suppression (Section 5.7).

A single COSY spectrum presents a map of the complete scalar coupling network in a macromolecular structure. Using the empirical rule that COSY cross peaks are with few exceptions observed only between protons separated by three or less covalent bonds in the amino acid structures (Table 2.2), Table 2.4 allows one to outline the regions a–h (Fig. 5.8C, below the diagonal) containing the connectivities between the following protons:

a. All nonlabile, nonaromatic amino acid side chain protons except βH–γCH$_3$ of Thr, δH–δH of Pro, and βH–βH of Ser.

b. αH–βCH$_3$ of Ala and βH–γCH$_3$ of Thr.

c. αH–βH of Val, Ile, Leu, Glu, Gln, Met, Pro, Arg, and Lys.

d. αH–βH of Cys, Asp, Asn, Phe, Tyr, His, and Trp.

e. αH–αH of Gly, αH–βH of Thr, δH–δH of Pro, αH–βH and βH–βH of Ser.

f. Aromatic ring protons, including the four-bond connectivity 2H–4H of His (Arseniev et al., 1982; King and Wright, 1982), and side chain amide protons of Asn and Gln.

g. Backbone NH–αH.

h. δCH$_2$–εNH of Arg.

The connectivities in the regions a–e and the cross peaks between aromatic protons in region f are most conveniently explored in ^{1}H COSY spectra recorded in D$_2$O solution after complete exchange of all labile protons. The side chain labile protons in regions f and h can only be observed in H$_2$O solution. Region g is empty after complete exchange of the labile protons in D$_2$O solution. In freshly prepared D$_2$O solutions of proteins it may contain cross peaks with slowly exchanging amide protons (Fig. 5.8B and C), and in H$_2$O solutions at pH below 6.0 it contains a complete set of backbone NH–αH cross peaks (Fig. 5.9A and B). Normally, the COSY spectrum of a diamagnetic protein contains no cross peaks outside the regions a–h. Exceptions may result from large conformation-dependent chemical shifts of individual resonance lines (Chapter 3). (The inhibitor K spectrum contains two such peaks, identified by arrows in Fig. 5.8C; additional weak peaks outside the regions a–h represent noise.) Large conformation shifts may also cause exchange of cross peaks between the spectral regions a–h. However, overall the reliability of this cross peak classification is better than 90%. (More specific identification of the cross peaks, including the relatively few peaks in unusual locations of the (ω_1,ω_2) plane, is obtained by the resonance assignment procedure described in Chapter 8.)

The backbone NH–αH region g in H$_2$O spectra is of special interest, since it presents a *fingerprint* of the amino acid sequence (Wagner and Wüthrich, 1982a). Each L-amino acid residue contributes a single NH–αH cross peak, and each Gly gives either two cross peaks, or a single cross peak with unique multiplet fine structure (Fig. 5.10). No cross peaks in region g come from the N-terminal residue, because of the rapid exchange of the α-amino protons, and from the prolines (Fig. 5.10). The COSY fingerprint of a protein (Fig. 5.9B) is a most useful guide through an NMR study. At the outset, compari-

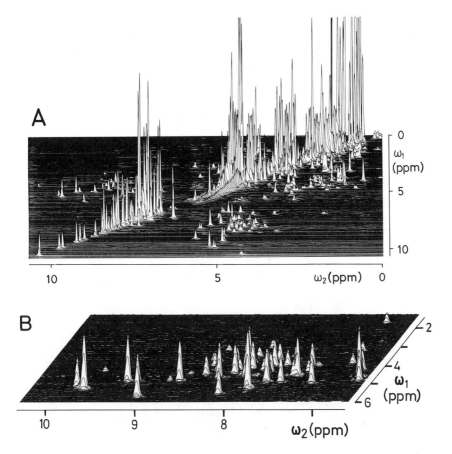

Figure 5.9. ¹H COSY spectrum of BPTI (0.02 M, H_2O, pH 4.6, 80°C; 500 MHz; absolute value). (A) Stacked plot of the entire spectrum. (B) Expanded presentation of the COSY fingerprint in the region (ω_1 = 1.8–6.0 ppm, ω_2 = 6.7–10.3 ppm) (from Wagner and Wüthrich, 1982a).

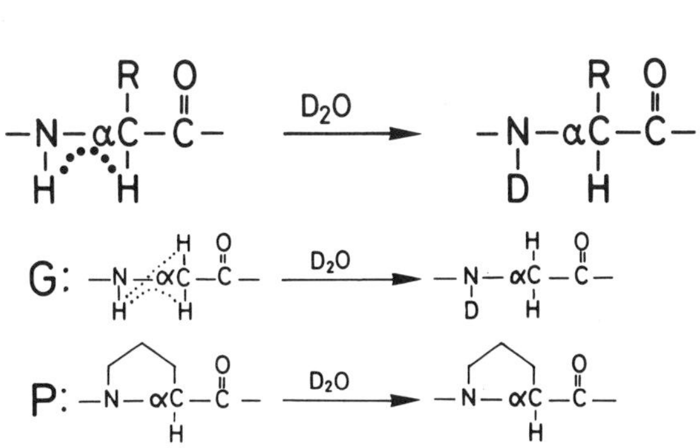

Figure 5.10. NH–αH connectivities (dotted lines) manifested in the COSY fingerprint of proteins and amid-proton exchange in D_2O solution.

son with the number of NH–αH cross peaks expected from the chemically determined amino acid sequence (Fig. 5.10) indicates whether all amino acid residues are represented by resolved cross peaks, and allows one an initial judgment of the homogeneity of the protein preparation. During an ongoing investigation, changes in the fingerprint may indicate deterioration of the protein (Štrop et al., 1983a). Further practical uses of the COSY fingerprint include studies on the molecular dynamics (Section 5.7) and checks on the sequence-specific resonance assignments (Chapter 8).

A third example (Fig. 5.11) shows the improved peak separation relative to the 1D NMR spectrum in the COSY fingerprint of the polypeptide–hormone glucagon bound in the lipid–water interphase of perdeuterated dodecylphosphocholine micelles (micelle-bound glucagon). The 1D spectrum between 7.3 and 9.0 ppm contains the resonances of the 28 backbone amide protons and two εNH lines of Arg, which are only partially resolved. In the COSY fingerprint, the cross peaks with amide proton resonances that over-

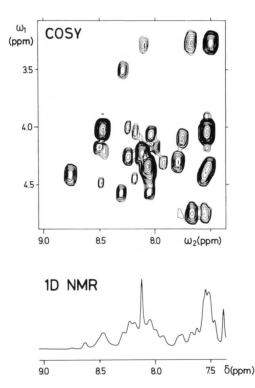

Figure 5.11. COSY fingerprint and 1D NMR spectrum of micelle-bound glucagon (0.015 M glucagon, 0.7 M [D$_{38}$]-dodecylphosphocholine, 0.05 M phosphate, H$_2$O, pH 6.0, 37°C; 360 MHz; absolute value). The COSY cross peak corresponding to the well separated NH line at 8.63 ppm in the 1D spectrum was bleached out by the water suppression (from Wider et al., 1982).

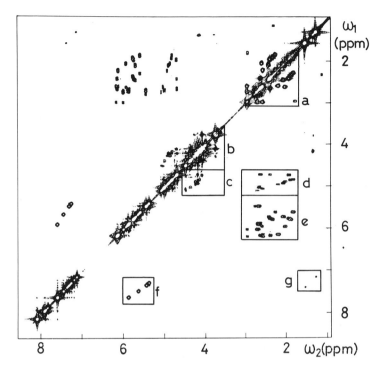

Figure 5.12. ¹H COSY spectrum of the DNA duplex d(CGCGAATTCGCG)₂ (0.002 *M* in duplex, 0.01 *M* phosphate, D₂O, pD 7.0, 37°C; 500 MHz; absolute value) (from Hare et al., 1983). Below the diagonal, solid lines define the regions a–g, which contain cross peaks between distinct types of protons (see text).

lap along ω_2 are separated thanks to the different shifts of the coupled α protons along the ω_1 axis. Generally, two COSY cross peaks coincide only when both *pairs* of coupled resonances have identical chemical shifts, which greatly reduces the probability of degeneracy compared to both 1D NMR and 2D-*J* spectroscopy.

Most of the general remarks on ¹H COSY spectra of proteins also apply to DNA fragments. However, all COSY spectra of nucleic acids reported so far were recorded in D₂O solution after complete exchange of all labile protons (Fig. 5.12). In contrast to proteins, this does not affect the information content, since no additional COSY cross peaks are expected from the presence of the labile protons (Table 2.5). Table 2.6 allows one to define the regions a–g (solid lines in the lower right triangle of Fig. 5.12) containing the following connectivities:

a. 2'H–2''H
b. 4'H–5'H, 4'H–5''H, and 5'H–5''H
c. 3'H–4'H.

d. 2'H–3'H and 2"H--3'H

e. 1'H–2'H and 1'H-2"H

f. 5H–6H of C

g. Four-bond connectivity 5CH₃–6H of T

Obviously, in all these regions the number of cross peaks can be compared with that expected from the nucleotide sequence. In particular, the regions f and g afford a straightforward check on the pyrimidine nucleotides present.

5.4. EXPERIMENTAL PROCEDURES

This section presents a list of procedures for optimizing the quality and appearance of macromolecular 2D NMR spectra. The list is unavoidably incomplete, since new methods and tricks are constantly being added. The selection of the techniques to be used and the practical execution depend on the problem to be solved, the spectrometer system, and to a large measure also on personal taste. For detailed instructions on individual procedures, the reader is referred to, for example, Wider et al. (1984), the textbooks by Bax (1982) and Ernst et al., (1987), and the manuals provided by the instrument manufacturers.

5.4.1. Absolute Value and Phase Sensitive Mode

The 2D FT of the time domain data set $s(t_1,t_2)$ can be considered as consisting of two subsequent 1D FT's, usually first with respect to t_2 and then with respect to t_1. Commonly, a cosine-FT and a sine-FT are performed along both time dimensions, which yields four data matrices

$$S^{cc}(\omega_1,\omega_2) \qquad S^{cs}(\omega_1,\omega_2) \qquad S^{sc}(\omega_1,\omega_2) \qquad S^{ss}(\omega_1,\omega_2) \qquad (5.1)$$

where, for example, $S^{cs}(\omega_1,\omega_2)$ is the frequency domain data matrix obtained through cosine-FT along t_1 and sine-FT along t_2. Pure 2D *absorption* signals can be obtained as linear combinations of the four data matrices in Eq. (5.1) (Bachmann et al., 1977). Up to 1983 phase sensitive spectra with big data matrices were difficult to obtain because of certain limitations of the commercial hardware, and therefore the 2D NMR spectra of biopolymers were presented in the absolute value mode

$$|S(\omega_1,\omega_2)| = [S^{cc}(\omega_1,\omega_2)^2 + S^{cs}(\omega_1,\omega_2)^2 + S^{sc}(\omega_1,\omega_2)^2 + S^{ss}(\omega_1,\omega_2)^2]^{1/2} \quad (5.2)$$

More recently, improved hardware was introduced and two different procedures for obtaining phase sensitive spectra from large data matrices, which may also be recorded with quadrature detection, were implemented (Marion and Wüthrich, 1983; States et al., 1982). Presently the phase sensitive mode is used almost exclusively. The main advantages of absorption spectra are the higher resolution and the possibility of distinguishing between positive and negative intensities. (For a Lorentzian line the broadening in the absolute value mode relative to the absorption mode is a factor $\sqrt{3}$ at half-height, and a factor 10 near the base line.)

5.4.2. Selection of the Experimental Scheme

A large variety of 2D NMR experiments, which can all be accomodated in the general scheme of Figure 5.4A, has already been described and there is a steady flow of new additions. Only a small selection of standard experiments are used in this book. [For example, variations of COSY include SECSY, SUPER COSY (Anil-Kumar et al., 1984), ω_1-decoupled COSY (Bax and Freeman, 1981; Rance et al., 1984), ω_1-scaled COSY (Brown, 1984; Frey et al., 1985b; Hosur et al., 1985a,c), COSY with variable flip angles $\neq 90°$ (Aue et al., 1976; Bax and Freeman, 1981), E. COSY (Griesinger et al., 1985), TOCSY (Braunschweiler and Ernst, 1983; Bax and Davis, 1985).] The selection of the experiment may be influenced by the available instrumentation (e.g., SECSY uses less memory space than COSY) or by the problem to be solved (e.g., ω_1-scaled COSY for systems where the relative chemical shifts of the different resonances are large compared to the spin–spin coupling constants). In this context it is also important to stress that pulse sequences such as those in Figure 5.4B to E, do not describe the complete 2D NMR experiments. Quite generally, in experimental schemes using nonselective pulses, different transfer processes for the coherence created at the end of the preparation period can occur simultaneously (Bain, 1984; Bodenhausen et al., 1984a). The results obtained with a particular pulse sequence then depend largely on proper selection of specified transfer pathways and suppression of unwanted peaks and artifacts, using techniques such as those described in Sections 5.4.3 to 5.4.5.

5.4.3. Phase Cycling

Phase cycling in 2D NMR experiments makes use of the fact that the coherence resulting from different processes occurring during evolution and mixing may be differently affected by variation of the phases of the rf pulses. By way of an example we consider the suppression of *axial peaks* in COSY. Axial peaks originate from z magnetization present immediately before the mixing pulse. (Two sources for z magnetization at the end of the evolution period are deviations of the first pulse from 90°, leaving residual z magnetization, and buildup due to T_1 relaxation during the evolution period.) Similar to a 1D spectrum the axial peaks are located on the ω_2 axis. While the COSY diagonal peaks and cross peaks are invariant under a 180° phase shift of the mixing pulse, the axial peaks change sign. Therefore, the following cycle of two experiments with different phases

$$90° \quad 90° \quad \text{Memory}$$

$$x \quad x \quad + \tag{5.3}$$

$$x \quad -x \quad +$$

where + indicates addition of the signal in the memory, will eliminate axial peaks. Two classical phase cycles are EXORCYCLE for combined suppression of axial peaks and separation of N and P peaks (Bodenhausen et al., 1977), and CYCLOPS for compensation of phase and amplitude errors in quadrature detection (Hoult and Richards, 1975). Phase cycling schemes for combined suppression of artifacts and selection of desirable coherence transfer pathways in 2D NMR may typically include 16,24,32, or even more steps. For work with proteins and nucleic acids this is no disadvantage, since the phase cycling can be combined with the indispensable accumulation of multiple scans for improvement of the S/N.

5.4.4. Purging, Multiple-Quantum Filters and z Filters

Purging techniques deal with selective suppression of undesired signals prior to data acquisition. Thereby, use is made of certain phase properties of these signals. Most purging techniques convert observable single quantum transitions into unobservable multiple quantum (MQ) coherence (Sørensen, 1984). With the use of multiple quantum filters (MQF) immediately before the detection period (Piantini et al., 1982; Rance et al., 1983; Shaka and Freeman, 1983) the spectra are purged from undesired features by selecting coherence transfer that proceeds through a certain order of MQ coherence. z filters (Sørensen et al., 1984) are also inserted before the detection period, and are used mainly to eliminate phase and multiplet distortions in the resulting 2D NMR spectra.

5.4.5. Suppression of Diagonal Peaks

2D NMR spectra of biopolymers are usually very crowded near the diagonal, and the strong diagonal peaks tend to mask nearby cross peaks (Figs. 5.8B, 5.9A, and 5.12). This situation can be improved through suppression of the diagonal peaks by recording difference spectra between two or several suitably modified experiments (Bodenhausen and Ernst, 1982; Nagayama et al., 1983). Selected diagonal peaks in COSY can also be suppressed with the use of MQF's (Piantini et al., 1982; Shaka and Freeman, 1983).

5.4.6. t_1 Noise, t_1 Ridges, and t_2 Ridges

Severe limitations for the analysis of 2D NMR spectra may arise from t_1 noise, that is bands of spurious signals running parallel to the ω_1 axis at the

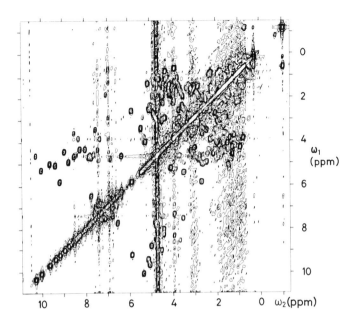

Figure 5.13. t_1 noise in the ^{1}H COSY spectrum of inhibitor K (same sample and conditions as in Fig. 5.8, except that the present spectrum was not symmetrized) (from Baumann et al., 1981).

positions of strong, sharp diagonal peaks (Fig. 5.13). Because of the important practical consequences, this phenomenon has attracted considerable interest (Mehlkopf et al., 1984; Nagayama et al., 1978), and several schemes for reducing t_1 noise either before or after the 2D FT are available (Baumann et al., 1981; Denk et al., 1985; Klevit, 1985). Recently it was recognized that in certain 2D NMR experiments, for example, NOESY and MQ spectroscopy, these vertical bands of spurious signals can contain t_1 ridges. These result from systematic errors in the *discrete* FT rather than from spectrometer instabilities or other sources of thermal noise, and can be eliminated by proper data handling during the 2D FT (Otting et al., 1986).

The appearance of t_2 ridges parallel to the ω_2 axis is another undesirable feature that may complicate the analysis of 2D NMR spectra. t_2 ridges can again originate from systematic errors in the discrete 2D FT, and may further include contributions from *truncation effects*. In practice, these t_2 ridges often have a negative sign and cause distortions of the base plane. They can be largely suppressed by empirical correction terms inserted during the 2D FT (Otting et al., 1986). [As an alternative, it has also been suggested to suppress the t_2 ridges through a maximum entropy treatment of the time domain data matrix, which would replace the 2D FT (e.g., Hore, 1985; Laue et al., 1985). This seems, however, hardly practicable with big data matrices in the near future.]

5.4.7. Performance Parameters, Spectral Resolution, and Sensitivity

Fundamentally, one wants to select the spectral ranges for ω_1 and ω_2 sufficiently large to include all resonance peaks. Otherwise peaks outside the frequency ranges covered may be folded back into the spectrum (e.g., Wider et al., 1981; FOCSY procedures may enable the use of reduced spectral ranges, see Wider et al., 1984). In NOESY and COSY the spectral ranges in ω_1 and ω_2 are the same. In a SECSY experiment $\omega_{1\,max}$ is equal to half the largest frequency difference between coupled nuclei. In delayed acquisition ^{1}H 2D-J spectra a value for $\omega_{1\,max}$ of ± 20 Hz to ± 30 Hz is normally sufficient, if strong coupling is neglected (Wider et al., 1981). Equally important is the selection of the number X_1 of equidistant t_1 values separated by the increment Δt_1. The value of Δt_1 is determined by the maximum frequency range $\omega_{1\,max}$ through the sampling theorem

$$\Delta t_1 \leq 1/(2\omega_{1\,max}) \tag{5.4}$$

X_1 determines the resolution in ω_1 and the total performance time. A large X_1 can at the same time increase the performance time and reduce the S/N, when the sampling process is extended too far into the decay of the signal along t_1. Along ω_2 the number of sampling points X_2 is usually limited only by the storage capacity. (When more data points are acquired along t_2, the relaxation delay during the preparation can be correspondingly shorter, so that the time required for the experiment is not noticeably affected by the choice of X_2.) Overall, the important message is that the expense in measuring time and reduced sensitivity for obtaining high resolution along ω_1 may easily become forbidding when working with macromolecules, whereas the resolution along ω_2 is usually limited only by the effective transverse relaxation time T_2^* in the system of interest. A typical set of performance parameters for a ^{1}H COSY experiment recorded at 500 MHz with the carrier frequency set to one end of the spectrum would be $\pm\ \omega_{1\,max}/2\pi = 5000$ Hz, $\omega_{2\,max}/2\pi = 10,000$ Hz, $X_1 = 512$, $X_2 = 8192$. The resulting resolution is 20 Hz/point in ω_1 and 2.5 Hz/point in ω_2. Further improvement of the resolution can be achieved by placing the carrier frequency in the center of the spectrum (Marion and Wüthrich, 1983; States et al., 1982; Wider et al., 1984) and by judicious choice of the experiment used (e.g., SECSY or ω_1-scaled COSY instead of COSY). As is further discussed in Section 5.4.9, the *digital resolution* is also routinely improved by zero filling, so that one would, for example, not work with a resolution of 20 Hz/point.

5.4.8. Resolution Enhancement and Base Line Correction

The appearance of 2D NMR spectra obtained from recordings with a given set of performance parameters can be improved by multiplication of the time domain data set $s(t_1,t_2)$ with suitable weighting functions $h_2(t_2)$ and $h_1(t_1)$. Base line correction and resolution enhancement are routinely applied.

Much work was devoted to optimizing weighting functions for resolution enhancement, such as the Lorentz–Gauss transformation, pseudoechos, and convolution difference methods (e.g., Bax et al., 1979; Campbell et al., 1973; Ernst, 1966). For 2D work with biopolymers the sine bell (DeMarco and Wüthrich, 1976) or, in improved versions, a shifted sine bell (Guéron, 1978; Wagner et al., 1978) or a squared sine bell are easily generated and applied. The effects of the shifted sine bell in Eq. (5.5)

$$\sin\left(\frac{\pi t}{t_2} + 0.04\pi\right) \tag{5.5}$$

on the FID and the 1D frequency spectrum of BPTI (Fig. 5.14) are representative for the common resolution enhancement routines. In the FID the signals at longer times in the detection interval are emphasized relative to those at the start of detection, which corresponds to suppression of the broad components relative to the sharp lines in the frequency spectrum. Evidently, improved resolution is obtained at the expense of reduced S/N, and therefore one has to compromise, for example, by judicious choice of the phase shift in Eq. (5.5). Rather strong resolution enhancement is indispensable for the presentation of absolute value 2D NMR spectra of macromolecules, as is exemplified by the NOESY spectrum of micelle-bound glu-

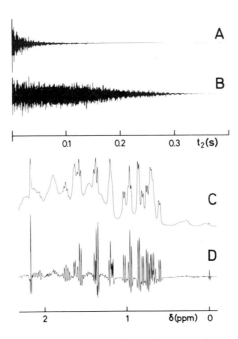

Figure 5.14. Resolution enhancement by digital filtering before FT. *(A)* and *(B)* 1D FID of BPTI before and after multiplication with the shifted sine bell of Eq. (5.5). *(C)* and *(D)* High-field region of the 1D ^1H NMR spectra obtained by FT of *(A)* and *(B)*, respectively.

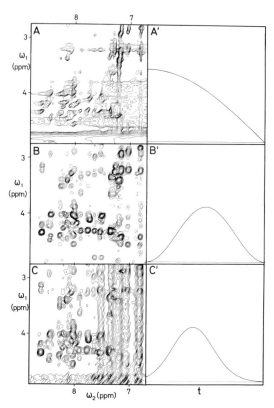

Figure 5.15. Improvement of an absolute value 2D ^{1}H NMR spectrum by different resolution enhancement functions. A region from a 500 MHz ^{1}H NOESY spectrum of the sample of micelle-bound glucagon in Figure 5.11 is shown as obtained after digital filtering in both dimensions with (A) the cosine function A', (B) the sine squared function B', and (C) the Gaussian function C' (from Wider et al., 1984).

cagon in Figure 5.15. After multiplication with a cosine function along t_1 and t_2, which produces no resolution enhancement, the broad tails of the absolute value peaks mask the information content of the spectrum (Fig. 5.15A). Of the two resolution enhancement functions of Figure 5.15B' and C', the squared sine bell was superior to the Lorentz–Gauss transformation, which corresponds closely to a phase shifted, squared sine bell.

5.4.9. Zero Filling

The visual appearance of the 2D NMR spectra resulting from the necessarily limited size of the time domain data matrix $s(t_1,t_2)$ can be improved by zero filling prior to FT (Bartholdi and Ernst, 1973). Particularly in the t_1 domain, zero filling is required in almost all cases. Experience gained with biopoly-

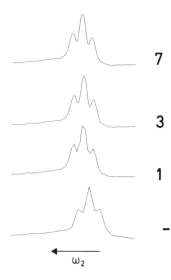

7

3

1

–

ω_2

Figure 5.16. Influence of zero filling in the time domain on the appearance of a multiplet along ω_2 in the frequency domain spectrum. The numbers on the right indicate the ratio of the number of points added by zero filling relative to the number of experimental data points (from Wider et al., 1984).

mer 2D ^{1}H NMR spectra showed that a noticeable improvement of the resolution can be achieved if N data points are supplemented by up to $3N$ zero values (Fig. 5.16).

5.4.10. Symmetrization

Ideal COSY or NOESY spectra recorded with identical performance parameters along t_1 and t_2 would be symmetrical with respect to the main diagonal. In practice, however, noise and instrumental artifacts tend to destroy the symmetry (Fig. 5.13). By a suitable combination of the upper left and the lower right triangle of the frequency spectrum obtained after FT, it is therefore possible to improve the significance of a 2D spectrum. In the procedure of symmetrization, the lower one of each pair of symmetrically located values is retained (Baumann et al., 1981).

Figures 5.13 and 5.8C show the same COSY spectrum obtained without and with symmetrization. (The spectra of Figures 5.8, 5.9, and 5.12 have all been symmetrized.) Symmetrization can lead to erroneous peaks when the original spectrum contains bands of spurious signals, since the crossing of two noise bands in the symmetrization process will feign the presence of two symmetrically located peaks. Such artifactual peaks arising from the t_1 noise in Figure 5.13 can be seen in the regions ($\omega_1 = 6.3$–7.6 ppm, $\omega_2 = 0.5$–2.5 ppm) and ($\omega_1 = 0.5$–2.5 ppm, $\omega_2 = 6.3$–7.6 ppm) of Figure 5.8C. Overall, it is advisable to rely on the combined use of symmetrized and original spectra for the data analysis. Particular care must be exercised when symmetrizing spectra recorded with markedly different performance parameters and data manipulations along t_1 and t_2.

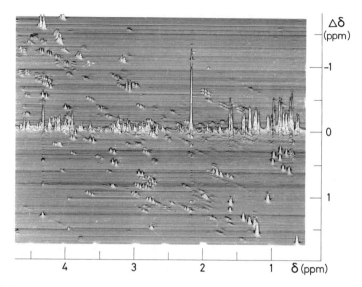

Figure 5.17. SECSY spectrum of BPTI (0.01 *M*, D$_2$O, pD 7.0, 68°C; 360 MHz; absolute value). Except for the aromatic region and some low field shifted αH lines; the complete spectrum of the nonlabile protons is shown (from Nagayama et al., 1980).

5.5. SCALAR (THROUGH-BOND) ^{1}H–^{1}H CONNECTIVITIES BY 2D NMR

During the relatively short period since 2D NMR experiments were first applied for studies of scalar coupling connectivities in proteins (Nagayama et al., 1979; 1980) and nucleic acids (Feigon et al., 1982), these techniques have been greatly improved. At the outset, the small data-storage capacity of the NMR spectrometers made SECSY spectra (Fig. 5.17; see also Fig. 5.6) attractive, since these could be recorded at higher digital resolution than COSY spectra. In the meantime, the following experiments have considerably extended the potentialities that used to be available with SECSY alone.

5.5.1. Phase-Sensitive COSY

Figure 5.18 illustrates with a 2D data set of rabbit liver metallothionein-2 (metallothionein-2) the improvement of the spectral resolution that can be achieved with otherwise identical data recording and manipulation by the phase sensitive mode when compared to the absolute value mode. In the absolute value spectrum (Fig. 5.18*B*) one recognizes outlines of the cross peaks with only partly resolved multiplet fine structure along ω_2. The pure *absorption mode* cross peaks (Fig. 5.18*A*) show well resolved multiplet fine structure in both frequency dimensions. The improved resolution is a consequence both of the narrower absorption lines and the *antiphase* intensity distribution in the absorption mode multiplets (inset in Fig. 5.18*A*). In practice the antiphase fine-structure patterns often enable unambiguous identifi-

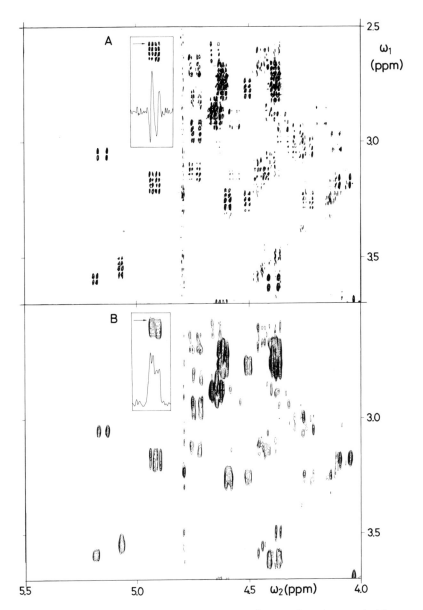

Figure 5.18. Phase-sensitive COSY. The same time-domain data set recorded for metal-lothionein-2 [0.005 M, D_2O, pD 7.0, 24°C; 500 MHz; carrier frequency in the center, $\omega_{1\,max}$ 2600 Hz, $\omega_{2\,max}$ 2600 Hz, X_1 512, X_2 2048; digital resolution after onefold zero filling 4.1 Hz/point in ω_1 and 1.0 Hz/point in ω_2; sine bell resolution enhancement with phase shifts of $\pi/32$ in t_1 and $\pi/64$ in t_2] was transformed (A) in the phase sensitive mode and (B) in the absolute value mode. A region containing $\alpha H-\beta H$ cross peaks is shown. The insets contain cross sections along ω_2 through the higher field $\alpha H-\beta H$ cross peak of Asp-2 taken at the ω_1 position indicated by the arrow. In the contour plots positive and negative levels are not distinguished.

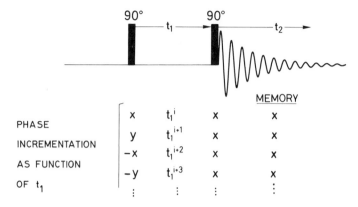

Figure 5.19. Time proportional phase incrementation scheme (TPPI) for recording phase sensitive, homonuclear COSY spectra with the carrier in the center of the spectrum. To end up with a phase-sensitive display in four quadrants, the phase of the first pulse is incremented by 90° in parallel with each incrementation of t_1, whereas the phase of the second pulse and the receiver phase are kept constant (from Marion and Wüthrich, 1983).

cation of cross peaks in difficult situations, for example, in noisy regions, near the diagonal, or in areas crowded with mutually overlapping cross peaks. To record the data of Fig. 5.18 with the performance parameters given in the figure caption, the *time-proportional phase incrementation* scheme (TPPI) (Bodenhausen et al., 1980; Marion and Wüthrich, 1983; Redfield and Kunz, 1975) was used. This scheme (Fig. 5.19) may be amended with other phase cycling routines for selection of particular coherence transfer pathways and for suppression of artifacts.

5.5.2. Multiple Quantum-Filtered COSY

Phase-sensitive COSY spectra can be considerably improved with the use of MQF's. So far, two-quantum filters (2QF) have been most widely used (e.g., Neuhaus et al., 1985; Rance et al., 1983), and 3QF-COSY and 4QF-COSY experiments with proteins were also described (Boyd et al., 1985b; Müller et al., 1985, 1986). Quite generally, with the availability of commercial spectrometers equipped with *digital phase shifters* (e.g., Hintermann et al., 1982), increased use of higher-order MQF-COSY experiments can be anticipated, and MQF's might also be implemented with 1D NMR experiments (Rance et al., 1985c). Based on the selection rules for MQF's (Braunschweiler et al., 1983; Müller et al., 1985), this opens interesting perspectives for editing complex spectra into subspectra devoid of spin systems with up to two, three, four, or five protons, which might considerably extend the scope of the spectral analysis.

In a conventional COSY experiment, the diagonal peaks are in pure dispersion when the phase is adjusted for pure absorption lines of the cross peaks. As a consequence, the pronounced tails of the intense diagonal dispersion lines mask the cross peaks near the diagonal. Figure 5.20A shows that in BPTI particularly extensive spectral regions are thus lost near the intense diagonal peaks of the methyl protons between 0.5 and 1.5 ppm, the methionine ε-methyl singlet at 2.13 ppm and the HDO singlet at 4.65 ppm. Routing of the coherence transfer through a 2QF (Fig. 5.21) allows one to obtain antiphase absorption modes for the cross peaks, and the diagonal peaks (except that in systems with more than two spins the diagonal peaks contain some dispersive antiphase contributions). A further improvement is obtained since the 2QF suppresses diagonal singlet peaks (Piantini et al., 1982; Shaka and Freeman, 1983). In Figure 5.20B the methionine methyl singlet and the solvent resonance were nearly completely suppressed. The basic four-step phase cycle for 2QF-COSY (Fig. 5.21) can be amended with other routines, for example, for quadrature detection in ω_1 (Fig. 5.19).

5.5.3. Relayed Coherence-Transfer Spectroscopy

A limitation in the use of COSY for studies of biopolymers arises in situations of accidental chemical shift degeneracy of resonance lines from different spin systems. For example, it happens quite frequently in proteins that the αH lines of different residues with otherwise well separated resonances are overlapped (Fig. 5.22). Even if all the COSY connectivities between vicinal protons in the scheme of Figure 5.22 are established, it will not be possible to identify the corresponding pairs of amide and β protons. Relayed coherence-transfer spectroscopy (RELAYED-COSY) allows one to overcome this fundamental problem, since two or several subsequent COSY steps can be performed in a single experiment (Bolton and Bodenhausen, 1982; Eich et al., 1982; King and Wright, 1983; Wagner, 1983). In this experiment, cross peaks are thus observed between spins i and k which are not directly coupled but share a mutual coupling partner j:

$$H_i(\omega_1) \xrightarrow{\text{COSY}} H_j \xrightarrow{\text{RELAY}} H_k(\omega_2) \tag{5.6}$$

In RELAYED-COSY the mixing pulse in the COSY scheme (Fig. 5.4D) is replaced by the sequence (90°–τ–180°–τ–90°), and in DOUBLE-RE-LAYED-COSY (which promotes one COSY step and two subsequent RE-LAY steps) by (90°–τ_1–180°–τ_1–90°–τ_2–180°–τ_2–90°), with phase cycling for selection of the desired transfer pathways and suppression of artifacts (Wagner, 1983). Compared to COSY, the delays τ, or τ_1 and τ_2, respectively, are new elements in these experimental schemes, which can be adjusted for optimal transfer via distinct values of the spin–spin coupling constants. For example, the efficiency of a relayed coherence transfer from NH to βH in the amino acid fragment NH–αCH–βCH is proportional to $\sin(\pi\ ^3J_{HN\alpha}\tau)\cdot$ $\sin(\pi\ ^3J_{\alpha\beta}\tau)$. An optimal relay transfer would thus be obtained if τ could be chosen to match both $1/(2\ ^3J_{HN\alpha})$ and $1/(2\ ^3J_{\alpha\beta})$. In work with biopolymers it is

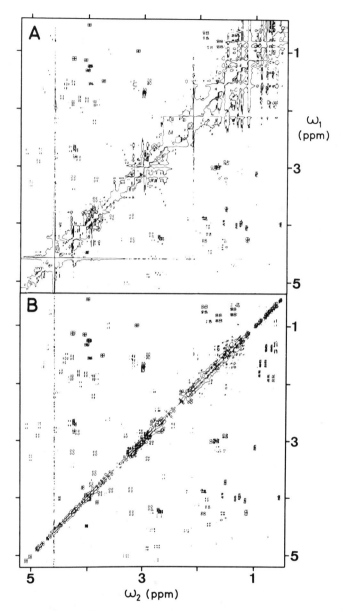

Figure 5.20. Comparison of conventional, phase-sensitive COSY *(A)* with phase-sensitive 2QF-COSY *(B)*. The region ($\omega_1 = 0.5–5.0$ ppm, $\omega_2 = 0.5–5.0$ ppm) of a spectrum of BPTI is shown (0.02 *M*, D_2O, pD 4.6, 36°C; 360 MHz; sine bell resolution enhancement with phase shifts of $\pi/4$ in t_1 and $\pi/8$ in t_2 applied in both spectra) (from Rance et al., 1983).

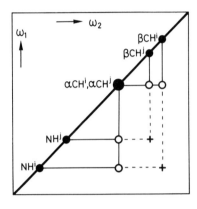

Figure 5.21. Experimental scheme for 2QF-COSY. τ must be sufficiently long to allow switching of rf phases between the second and the third pulse (τ was 4 μs in the experiment of Fig. 5.20B). The basic four-step phase cycle includes phase shifts for the third pulse, and for the data routing in the computer.

Figure 5.22. COSY and RELAYED-COSY connectivities for the NH–αCH–βCH fragments of two valines i and j, with identical αH chemical shifts. Filled circles indicate diagonal peaks, open circles are COSY cross peaks, and crosses are RELAYED-COSY cross peaks.

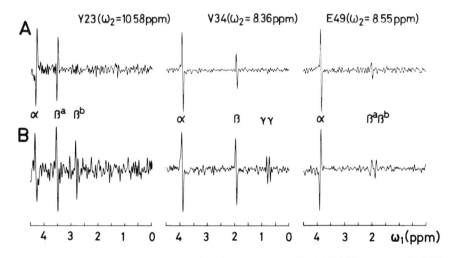

Figure 5.23. RELAYED-COSY (A) and DOUBLE-RELAYED-COSY (B) spectra of cBPTI (0.01 M, H$_2$O, pH 4.6, 36°C; 500 MHz; τ = 30 ms, and τ_1 = 25 ms, τ_2 = 31.2 ms, respectively). Cross sections along ω_1 at the ω_2 positions of the amide protons of Tyr-23, Val-34, and Glu-49 from phase-sensitive spectra are shown. The Greek letters identify the αH–NH, βH–NH, and γH–NH cross peaks, where $^3J_{\alpha\beta a}$ is larger than $^3J_{\alpha\beta b}$ (from Chazin et al., 1985).

usually advantageous either to aim for a compromise in the selection of the delays, so as to promote transfer for a wide range of coupling constants [thereby the influence of *passive* spin–spin couplings must also be considered (Bax and Drobny, 1985; Chazin et al., 1985, 1986; Weber et al., 1985a)] or to perform the same experiment repeatedly with different delay times. Figure 5.23 shows that in addition to the relayed peaks, a RELAYED-COSY spectrum also contains *direct* COSY connectivities, and a DOUBLE-RELAYED-COSY spectrum also contains direct and relayed connectivities. Usually distinction of the different connectivities follows readily from the different chemical shift ranges for the different proton types (Tables 2.4 and 2.6) and from comparison with the COSY spectrum. Furthermore, phase cycling schemes can be devised that produce different modes for the different connectivities, for example, pure absorption peaks for relayed connectivities, and mixed phases for diagonal peaks and direct connectivities. For Val-34 (Fig. 5.23), RELAYED-COSY yields the transfer $\beta H \rightarrow \alpha H \rightarrow NH$ and DOUBLE-RELAYED-COSY the pathways $\gamma^1 CH_3 \rightarrow \beta H \rightarrow \alpha H \rightarrow NH$ and $\gamma^2 CH_3 \rightarrow \beta H \rightarrow \alpha H \rightarrow NH$. In Tyr-23 and Glu-49 the relay pathway is selective for the β proton with the larger coupling constant $^3J_{\alpha\beta}$, and the double-relay transfer, $\beta^b H \rightarrow \beta^a H \rightarrow \alpha H \rightarrow NH$, is via the geminal coupling $^2J_{\beta\beta}$ (Fig. 5.23).

5.5.4. Multiple Quantum Spectroscopy

This is an alternative to the use of COSY and relayed coherence-transfer techniques, and it can also be used to extend the capabilities of these experiments in important ways. It can be foreseen that similar to the aforementioned use of MQF's, the scope of practical applications of MQ spectroscopy will in the near future be widened considerably by further improvements of commercial NMR spectrometers.

Fundamentally, the most exciting prospects for the use of MQ techniques arise from the fact that by selecting MQ coherence of order p, one should in principle be able to obtain subspectra containing exclusively resonances from spin systems with at least p spins. Thus, future applications might in principle include stepwise *editing* of protein ^1H NMR spectra into subspectra devoid of spin systems with two, three, four, and five protons by use of 3-, 4-, 5-, and 6-quantum spectroscopy, which could greatly facilitate the spectral analysis. Another important role for MQ spectroscopy arises because the peak intensities in COSY and RELAYED-COSY are not in general a faithful manifestation of the number of spins observed, so that it may become difficult to identify multiple equivalent nuclei and distinguish between spin systems such as AX and AX_2, or A_2X_2 and A_3X_3. In MQ spectra, *equivalence* of two or several spins is unambiguously recognized. Furthermore, *direct* connectivities seen in COSY and *remote* connectivities seen in RELAYED-COSY are also manifested in MQ spectra, which have an advantage over the other techniques because the absence of diagonal peaks facilitates observation of cross peaks between nuclei with similar chemical shifts.

A comprehensive survey of MQ spectroscopy for work with solutions was presented by Braunschweiler et al. (1983). Practical applications of zero quantum and 2Q techniques with biopolymers have been described (e.g., Boyd et al., 1983, 1985a; Macura et al., 1983; Otting and Wüthrich, 1986; Rance et al., 1985a; Wagner and Zuiderweg, 1983), and higher-order MQ experiments are pursued in several different laboratories. Practical limitations for MQ spectroscopy arise because the excitation of MQ transitions of a particular order is a priori not uniform. In a conventional MQ experiment (Fig. 5.24A) the interval τ is tuned to a preselected value for the spin–spin coupling constants in the system studied. Improvements over the conventional experiment can be achieved with symmetrical excitation–detection schemes (Fig. 5.24B), which enable uniform excitation over a wide range of spin–spin couplings, and with proper phase cycling can provide unambigu-

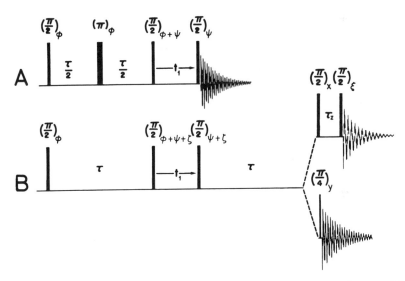

Figure 5.24. (A) Convential pulse sequence for MQ spectroscopy (note that $\pi = 180°$). For excitation of p quantum coherence, ϕ is cycled through $k\pi/p$, with $k = 0,1,2,$ . . . , $2p - 1$, and the FID's are alternately added and subtracted. ψ is 0 for even- and $\pi/2$ for odd-quantum excitation. (B) Example of a symmetrical excitation–detection scheme for uniform excitation of MQ coherence. Experiments recorded with different distinct values of τ are added up. ζ is cycled through the two values 0 and $\pi/2$, and the results of the two experiments are then subtracted. Removal of undesired magnetization components can be improved either by a $(\pi/4)_y$ purging pulse applied in all experiments of a sequence (lower trace) or by a z filter (upper trace). The last pulse of the z filter is phase cycled in steps of $\pi/2$ along with the receiver, and τ_z can be varied within a sequence of otherwise identical experiments to suppress zero quantum coherence present during τ_z (from Rance et al., 1985a).

ous criteria for distinguishing between direct and remote connectivities based on the sign of the absorption peaks.

Characteristic patterns for the geometric arrangement of the 2Q signals manifesting direct connectivities, remote connectivities, and equivalence in a network of coupled spins are exemplified in Figure 5.25 for alanine. ω_2 is the usual chemical shift axis and ω_1 represents the 2Q frequencies. The broken line is the 2Q skew diagonal ($\omega_1 = 2\omega_2$). Each of the direct connectivities NH $\rightarrow \alpha$H and αH $\rightarrow \beta$H is manifested by a pair of signals symmetrically disposed relative to the skew diagonal at the ω_2 positions of the two interacting spins [i.e., with 2Q frequencies equal to δ(NH) + $\delta(\alpha$H), and $\delta(\alpha$H) + $\delta(\beta$H), respectively]. The remote connectivity NH $\rightarrow \beta$H is represented by a lone peak at the ω_2 position of the common coupling partner αH. A horizontal line drawn through this signal disects the skew diagonal at an ω_2 position that coincides with the center between δ(NH) and $\delta(\beta$H) and does not match the chemical shift of any of the protons, and the 2Q frequency equals δ(NH) + $\delta(\beta$H). Equivalence of the β methyl protons is manifested by a lone peak at the ω_2 position of αH, which is the common coupling partner of the different equivalent spins. A horizontal line through this signal disects the skew diagonal at the ω_2 position of βH, with a 2Q frequency of $2\delta(\beta$H). At the position [$\omega_1 = 2\delta(\beta$H),$\omega_2 = \delta(\beta$H)] one usually observes a *forbidden* auto peak from the methyl protons (Müller et al., 1985).

The 2Q spectrum of the dinucleotide d(TpTp) in Figure 5.26 was recorded with the symmetric excitation scheme of Figure 5.24*B*. All direct and remote connectivities can be identified with the use of the indicated positions for the individual proton chemical shifts and the 2Q frequencies, except for the peaks in the areas a, b, and c and the direct connectivities in symmetrical

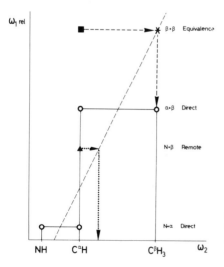

Figure 5.25. Manifestation of direct connectivities (O), remote connectivities (▲), and equivalence (■) in the ^{1}H 2Q spectrum of an alanyl residue, –NH–αCH(βCH$_3$)–CO–. The star indicates a "forbidden" autopeak of the methyl protons on the skew diagonal.

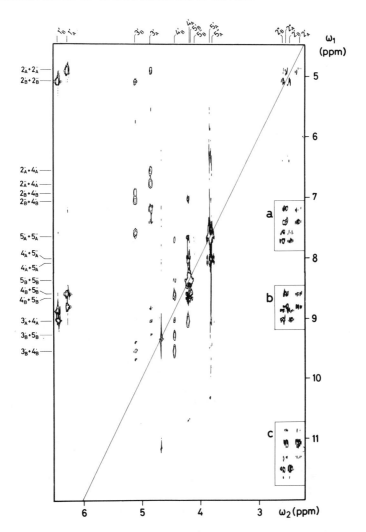

Figure 5.26. Phase-sensitive 2Q spectrum of d(TpTp) (0.005 M, D_2O, 0.05 M PO_4^{3-}, 0.1 M NaCl, pD 7.0, 35°C; 300 MHz; experiment of Figure 5.24B with purging pulse, 11 τ values evenly spaced between 40 and 240 ms). Chemical shifts and 2Q frequencies are indicated along ω_2 and ω_1, respectively, where A is the 5′-terminal and B the 3′-terminal nucleotide. The 2Q skew diagonal ($\omega_1 = 2\omega_2$) has been drawn in for reference, and both positive and negative contour levels are drawn. For the regions a, b, and c, see Figure 5.27 (from Rance et al., 1985a).

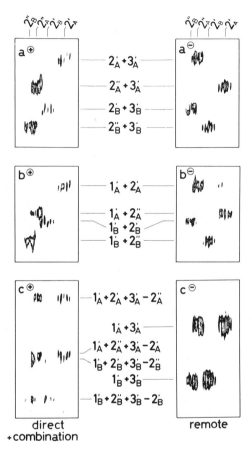

Figure 5.27. Edited subspectra for the regions a, b, and c in the 2Q spectrum of Figure 5.26, containing either exclusively positive or exclusively negative contour levels (from Rance et al., 1985a).

locations with respect to those in the areas a and b. For example, the topmost peaks correspond, from left to right, to 2Q coherence between the protons $2'_A$ and $2''_A$, which is transferred to single quantum coherence on $1'_A$ or on $3'_A$, or observed as a pair of direct connectivity peaks in symmetrical positions with respect to the skew diagonal. The advantage arising from the absence of diagonal peaks is readily apparent when inspecting the direct connectivities between 2'H and 2''H, or between 5'H and 5''H. Editing of the spectral regions a, b, and c into subspectra containing, respectively, direct connectivities and remote connectivities was achieved by plotting positive and negative contour levels separately (Fig. 5.27). In addition to a complete set of direct and remote peaks, the spectrum contains some *combination peaks* (Rance et al., 1985a); in the region c the 2Q frequencies of the combination peaks are also indicated.

5.6. CROSS PEAK FINE STRUCTURE AND MEASUREMENT OF CHEMICAL SHIFTS AND SPIN–SPIN COUPLING CONSTANTS

The resolved fine structure in phase-sensitive 2D NMR spectra (Figs. 5.18A, 5.20) can be of great help for obtaining resonance identifications and accurate measurements of NMR parameters, as is illustrated in Figure 5.28 with the first-order splitting patterns expected for four ^{1}H spin systems of amino acid residues in phase-sensitive COSY or 2QF-COSY spectra. In homonuclear correlation spectroscopy, signals along both ω_1 and ω_2 appear exclusively at frequencies that are also present in the 1D NMR spectrum. [This is because in COSY or NOESY (Fig. 5.4B and D) the behavior of the spins during t_1 can be manifested in the time-domain data only as phase or amplitude modulation of the normally observed spectral frequencies; these 2D pulse sequences cannot create new frequencies during t_2. In MQ spectros-

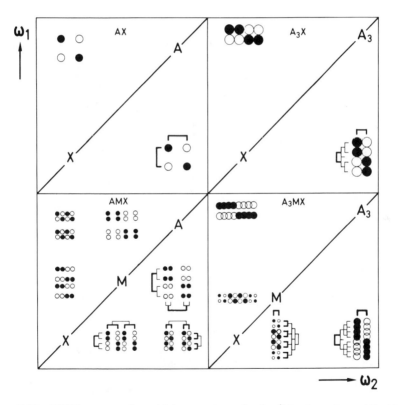

Figure 5.28. COSY cross peak multiplet structures for the ^{1}H spin systems AX, AMX, A_3X, and A_3MX. The letters on the diagonal indicate the chemical shifts for the different groups of protons. Filled circles represent positive signals, open circles negative signals, and the area of the circles represents the peak intensity. In the lower right of each spectrum, active and passive couplings are indicated with thick and thin lines, respectively (adapted from Neuhaus et al., 1985).

copy, linear combinations of the normal NMR frequencies can occur along ω_1.] For predicting the fine structure one must further take account of the antiphase character of COSY cross peaks (Fig. 5.18A). Thereby an important distinction must be made between the *active* coupling for a given cross peak, which in COSY is the coupling between the protons connected by the cross peak, and all other, *passive* couplings between these protons and any others. For a coupling network of the type H_i–H_a–H_b–H_j, where H_a and H_b are the spins connected by the cross peak, and H_i and H_j represent all the passive coupling partners of H_a and H_b, respectively, the splitting J_{ab} appears always in antiphase along both frequency axes, whereas the passive couplings J_{ai} and J_{bj} appear as additional in-phase splittings along either ω_1 or ω_2. (These relations can be derived using the product-operator formalism of Sørensen et al., 1983; see Neuhaus et al., 1985). As a consequence, for the coupling network H_i–H_a–H_b–H_j the components of a COSY cross peak at $(\omega_1 = \delta_a, \omega_2 = \delta_b)$ will appear as a rectangular array, cross sections along ω_1 having the multiplet frequencies of the H_a resonance, and cross sections along ω_2 having those of the H_b resonance. For the cross peak at $(\omega_1 = \delta_b, \omega_2 = \delta_a)$, the multiplet frequencies of H_a are along ω_2 and those of H_b along ω_1.

In the simplest possible case of an AX spin system of Gly (Fig. 5.28), there are no passive couplings. The active coupling produces a square array of two positive and two negative peaks, such that cross sections along either frequency axis show an antiphase splitting of magnitude J_{AX}. This *antiphase square array* is the basic unit from which all absorption-mode COSY cross peak multiplet patterns are built up. Within one properly phased COSY data set, all such antiphase square arrays have the same sign distribution (e.g., positive at top left and bottom right). This pattern can also be used for recognizing cross peaks in COSY spectra with automatic computer search procedures (Meier et al., 1984; Pfändler et al., 1985).

In AMX systems with nonvanishing spin–spin couplings J_{AM}, J_{AX}, and J_{MX} (Fig. 5.28), the AX pattern is split into four identical arrays separated by the passive couplings. For example, the MX cross peak at $(\omega_1 = \delta_X, \omega_2 = \delta_M)$ manifests along ω_1 the antiphase splitting J_{MX} and the in-phase splitting J_{AX}, and along ω_2 the antiphase splitting J_{MX} and the in-phase splitting J_{AM}. It is readily apparent from Figure 5.28 that the appearance of the cross peaks, including the order in which positive and negative components are arranged, depends critically on the relative size of the individual coupling constants. These considerations on AMX spin systems can be extended to systems with more than three nondegenerate spins $\frac{1}{2}$. For example, the NH–αH cross peak from a NH–αCH–βCH$_2$ fragment of Cys shown in Figure 5.29 displays the active coupling $J_{HN\alpha}$ along ω_2, and the active coupling $J_{HN\alpha}$ as well as the passive couplings $J_{\alpha\beta}$ and $J_{\alpha\beta'}$ along ω_1.

The treatment of spin systems involving groups of two or more magnetically equivalent protons can use the same concepts, provided that *each proton is formally treated separately*. For example, for an A_3X system (Fig.

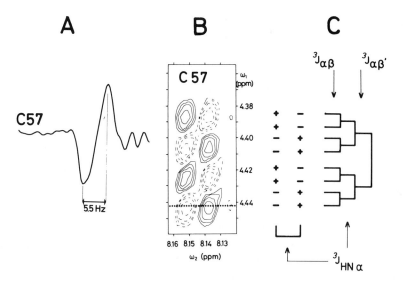

Figure 5.29. Amide-proton–α-proton cross peak fine structure for Cys-57 in a phase-sensitive COSY spectrum of BUSI. *(A)* Cross section taken along the dotted line in the contour plot. *(B)* Contour plot. Positive contours are drawn with solid lines, negative contours with broken lines. *(C)* Spin–spin coupling scheme (from Marion and Wüthrich, 1983).

5.28) the active coupling must then be taken to be that between X and only one of the three methyl protons, the remaining two protons A acting as passive coupling partners of X. Normalized superposition of three identical contributions, in each of which a different A proton is considered to be the active coupling partner of X, yields the total cross peak intensity. [This argument may be more rigorously justified in terms of the product-operator formalism of Sørensen et al. (1983); see Neuhaus et al. (1985).] In analogy to the familiar Pascal triangle describing the first-order in-phase splitting patterns resulting from spin–spin couplings with equivalent spins $\frac{1}{2}$ (Fig. 5.30A), an *antiphase Pascal triangle* may be constructed, in which the $(n + 1)$th row

```
A                              B
         1                              1
        1  1                           1 -1
       1  2  1                        1  0 -1
      1  3  3  1                     1  1 -1 -1
     1  4  6  4  1                  1  2  0 -2 -1
    1  5 10 10  5  1               1  3  2 -2 -3 -1
   1  6 15 20 15  6  1            1  4  5  0 -5 -4 -1
```

Figure 5.30. Pascal triangle *(A)* and its antiphase analog *(B)* (from Neuhaus et al., 1985).

represents the intensity pattern expected for the X resonance of an A_nX spin system as it appears in the absorption mode COSY cross peak connecting X and A (Fig. 5.30B). Note that in each row the J_{AX} splitting appears in antiphase only once, the remaining $(n-1)$ splittings being in phase. The fourth row of the antiphase analog of the Pascal triangle thus predicts for an A_3X spin system that sections through the cross peak along the multiplet structure of the X resonance show an intensity pattern of $1,1,-1,-1$ (Fig. 5.28), which coincides with the experimental observations (Fig. 5.31).

Figure 5.28 further shows the cross peak fine structures expected for an A_3MX spin system in which $J_{MX} = \frac{1}{2}J_{AX}$ and $J_{AM} = 0$ (this represents a Thr spin system, with A_3, M, and X representing γCH_3, αH, and βH, respectively). Here, the couplings of A_3 to X are active in the A_3X cross peaks and give the $1,1,-1,-1$ pattern just described, whereas in the MX cross peaks they are passive, resulting in the more familiar $1,3,3,1$ intensity pattern (Fig. 5.30A). In the A_3X cross peaks the passive coupling to the single proton M results in a further doubling of the $1,1,-1,-1$ pattern. In Figure 5.31 this additional splitting causes some line broadening, but it is not resolved. In the MX cross peaks, the $1,3,3,1$ fine structure is superimposed on the antiphase MX doublet splitting (Fig. 5.28).

In the spectra of biological macromolecules, the theoretical fine structure

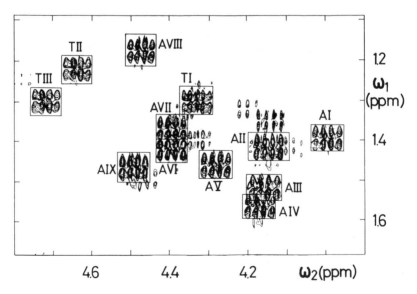

Figure 5.31. Spectral region containing the $\alpha H-\beta CH_3$ cross peaks of Ala and the $\beta H-\gamma CH_3$ cross peaks of Thr in a phase-sensitive 2QF-COSY spectrum of metallothionein-2 (same spectrum as Fig. 5.18). Positive and negative levels are plotted without distinction. The framed multiplets are identified by the amino acid type and an arbitrary Roman numeral index; additional, weaker peaks are due to the presence of a second metallothionein (from Neuhaus et al., 1985).

patterns are often complicated by the effects of finite linewidth and limited digital resolution. This must be taken into account when using fine structure features for resonance identifications or measurements of NMR parameters. Quite generally, when two Lorentzian lines overlap, the extrema of the resulting envelope do not correspond to the true centers of the original lines. Lines that are in phase tend to amalgamate, their combined envelope showing only one maximum. For lines in antiphase, overlap causes partial cancellation, thus reducing the overall intensity of the combined envelope, and the apparent line separation is larger than the true separation. For two identical lines in antiphase, a limiting separation of ∼ 0.576 times the linewidth is reached, beyond which the only effect of further reducing the true separation is to reduce the overall intensity of the combined envelope. These relations are summarized in Figure 5.32A–C. For multiplets with more than one splitting the situation is more complex, but can in principle be qualitatively assessed using these fundamental considerations on overlapping in-phase and antiphase lines.

Reduction of the digital resolution has similar effects on the appearance of fine structure patterns as an increase of the linewidth. Therefore, different apparent fine structures are usually observed for corresponding cross peaks

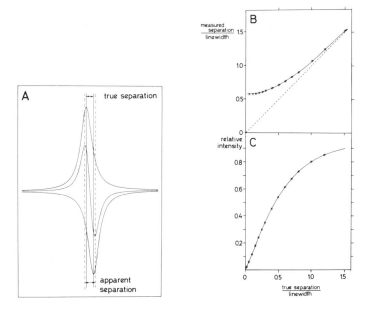

Figure 5.32. Effects of line broadening on the apparent line separation and the overall cross peak intensity in antiphase multiplets. *(A)* Definition of *true separation* and *apparent separation* in an antiphase doublet simulated by addition of two Lorentzian lines with opposite sign. *(B)* and *(C)* Plots versus the ratio of true separation over linewidth for the apparent experimental peak separation and for the observed relative intensity (from Neuhaus et al., 1985).

in the locations ($\omega_1 = \delta_A$, $\omega_2 = \delta_X$) and ($\omega_1 = \delta_X$, $\omega_2 = \delta_A$). In the AMX system of Figure 5.33, the components of the αH multiplet are more closely spaced than those of either βH multiplet (see also Fig. 5.28), so that the degree of internal cancellation within a cross peak is higher when the αH resonance appears along the poorly digitized ω_1 dimension. For this reason, the αH–βH cross peaks above the diagonal are more intense than the corresponding peaks below it. In the αH–β'H cross peak below the diagonal, particularly severe cancellation along ω_1 occurs for the inner two components, resulting in an apparent splitting of the remaining outer components that is much wider than the true coupling. Overall, the antiphase character of the cross peaks is the prime cause of the cross peak intensity variations in COSY spectra (Figs. 5.8, 5.9, 5.11, 5.12, 5.18), since reduction of peak intensity starts already at a ratio of true line separation to linewidth of approximately 2.0 (Fig. 5.32C).

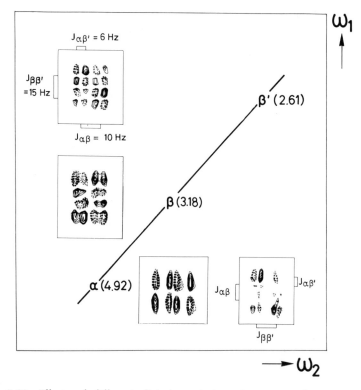

Figure 5.33. Effects of different digital resolution along ω_1 and ω_2 (4.9 and 1.0 Hz/point, respectively) on cross peaks in phase-sensitive 2QF-COSY spectra. The αH–βH cross peaks of Asp-2 in metallothionein-2 (same spectrum as Fig. 5.18) are shown, with the chemical shifts indicated along the (noncontinuous) diagonal. Positive and negative levels are plotted with broken and solid lines, respectively. The size of the coupling constants is indicated in the upper left (from Neuhaus et al., 1985).

These qualitative considerations on the influence of linewidth and digitization show that the position of the central node in cross sections along ω_1 or ω_2 through an absorption mode, antiphase COSY cross peak is invariant to resolution effects, and consequently is an excellent measure of chemical shifts. Spin–spin coupling constants can be extracted from the same spectra by measurements of the multiplet peak separation along ω_2 (Fig. 5.29). However, splittings observed in antiphase multiplet structures are an unreliable measure of coupling constants whenever these are appreciably smaller than the linewidth (Fig. 5.32) (Neuhaus et al., 1985; Oshkinat and Freeman, 1984). As a rule of thumb, systematic errors in J values thus measured in small proteins and nucleic acid fragments must be expected for $|J| \leq 5$ Hz.

Additional complications may arise from the prevalence of strong coupling. As an illustration, Figure 5.34 shows the αH–βH cross peaks from three AMX spin systems. Some qualitative conclusions may be drawn from visual inspection of these patterns. For example, the symmetrical appearance of the multiplet structures resulting from the overlap of the two cross peaks αH–βH and αH–β'H in Cys-57 indicates that the two active couplings $^3J_{\alpha\beta}$ and $^3J_{\alpha\beta'}$ are nearly identical. However, reliable analysis and measurement of chemical shifts and spin–spin coupling constants will usually require comparison with spectrum simulations (Fig. 5.34). In practical applications of such spectrum simulations, it is advisable to analyze both of the symmetry related cross peaks above and below the diagonal.

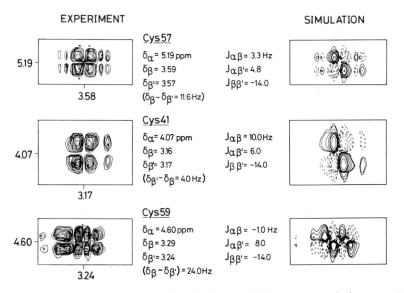

Figure 5.34. Strong spin–spin coupling in the multiplet patterns of phase-sensitive COSY spectra. Three αH–βH cross peaks of Cys in metallothionein-2 (same spectrum as Fig. 5.18) are compared with spectral simulations computed with the program SPHINX, using the parameters indicated in the figure. In the simulations, positive and negative peaks are drawn with solid and broken lines, respectively.

5.7. STUDIES OF LABILE PROTONS WITH 2D NMR

To prevent loss of cross peaks with labile protons, 2D NMR spectra must often be recorded in H_2O solution. The spectroscopist then faces serious experimental problems, since the weak solute signals (typical concentration range 0.001–0.01 M) have to be detected in the presence of the huge signal from the water protons (concentration 110 M). Similar to the situation with 1D NMR experiments (Section 3.4), different 2D NMR procedures are employed for experiments with slowly exchanging interior amide protons in proteins, the more rapidly exchanging solvent-accessible amide protons at slightly acidic pH, or the highly labile protons in nucleic acids or in polypeptides at high pH (Section 2.3).

Interior amide protons in proteins can often be observed in D_2O, so that no special precautions are needed for suppression of the solvent line (Figs. 5.8 and 5.13). In this situation the superior resolution of 2D NMR spectra can be exploited for obtaining more complete data on the kinetics of the proton exchange (Fig. 5.10) than with 1D NMR (Fig. 3.13). Figure 5.35A shows the COSY fingerprint of BPTI (Fig. 5.9B) at different times after the protein was dissolved in D_2O. In these experiments the proton exchange during the NMR measurements (a single COSY spectrum was recorded in 12 h) was quenched by lowering the temperature. Exchange rates slower than approximately 0.1 min^{-1} at 36°C can thus be measured. For example, the exchange rates obtained from the cross sections in Figure 5.35B are 1.9 × 10^{-3}, 2.8 × 10^{-4}, 4.5 × 10^{-2}, 1.1 × 10^{-5}, and 4.9 × 10^{-4} min^{-1} for Cys-5, Glu-7, H Ala-27, Leu-29, and Thr-54, respectively. Instead of COSY, NOESY can also be employed for such exchange studies (Boelens et al., 1985).

The amide protons in H_2O solutions of proteins can be studied using selective saturation of the solvent line (Fig. 5.9). Five experimental schemes, which have all been used in practice for both COSY and NOESY, are shown in Figure 5.36. In these schemes, time shared irradiation during t_2 (A and C) tends to produce spurious signals in the spectra, and continuous irradiation during t_1 (B and C) can cause *Bloch–Siegert shifts* for resonance lines near the solvent. However, since the t_1 values employed for work with biological macromolecules are relatively short, satisfactory suppression can be achieved with the experiments D or E, which are therefore in most cases the best choice. Figure 5.37 shows a representative spectrum recorded with the suppression scheme D. The major residual perturbation is a band of t_1 noise at the ω_2 position of the solvent line (which can be removed by symmetrization; see Fig. 5.9A and Wider et al., 1984). Furthermore, the cross peaks with α protons that overlap with the solvent line are *bleached out* (Anil-Kumar et al., 1980b), so that a narrow spectral region of approximately ± 0.05 ppm about the solvent position is empty. Typically, up to five cross peaks are thus lost in the COSY fingerprint of a small protein (e.g., Fig. 5.11). A complete data set can usually be obtained by a combination of measurements at two different temperatures. An alternative is provided by

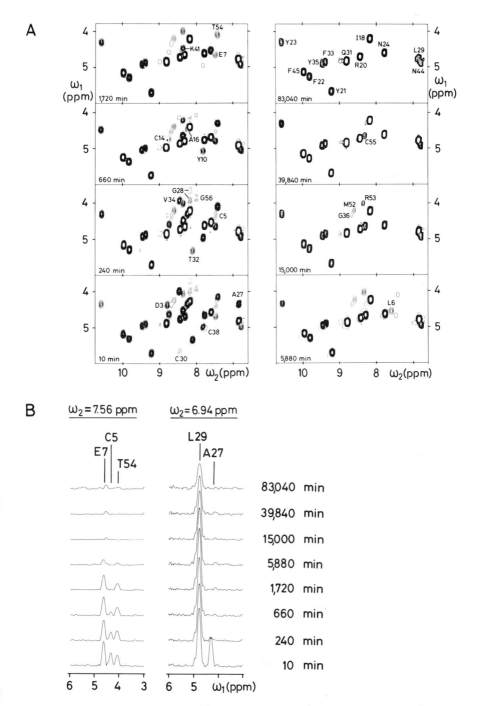

Figure 5.35. Amide-proton exchange observed in the ^1H COSY spectrum of a D_2O solution of BPTI (0.02 M, pD 3.5; 500 MHz; absolute value). The protein solution was prepared at 24°C, then heated at 36°C during the period indicated with the individual spectra, and subsequently measured at 24°C. (A) Contour plots. The exchanging protons are identified in the last spectrum where they can be seen. (B) Two cross sections through (A) along ω_1 (from Wagner and Wüthrich, 1982b).

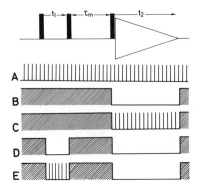

Figure 5.36. Solvent suppression by saturation in H_2O solutions of proteins. The traces A–E show five different switching cycles for the selective irradiation of the water resonance during a NOESY experiment (top trace). A shaded bar indicates continuous irradiation, vertical lines represent time shared irradiation, and empty spaces indicate that the decoupler was turned off (from Wider et al., 1983).

the direct connectivities in a 2Q spectrum recorded with solvent saturation, which represent a complete fingerprint (Otting and Wüthrich, 1986). For completeness sake it should be added that H_2O spectra are recorded in a mixture of H_2O and D_2O (usually 9:1), where the D_2O provides the lock signal, and that the quality of the spectra obtained with solvent saturation in H_2O, in particular the width of the bleached-out region, depends critically on optimal shimming of the magnetic field H_0 and tuning of the probe.

Solvent suppression in H_2O solutions of nucleic acids, or polypeptides under conditions of rapid exchange, can be achieved by use of semiselective pulses (Fig. 5.2) in the 2D NMR experimental schemes. For example, with the use of a time shared long pulse (Fig. 5.2C) as the observation pulse in the NOESY scheme (Fig. 5.4B), the solvent resonance can be suppressed by a

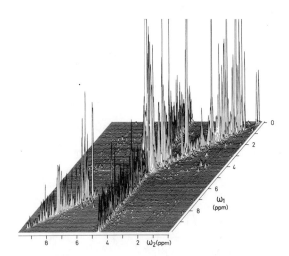

Figure 5.37. Solvent suppression by selective saturation (Fig. 5.36D) in the ^1H NOESY spectrum of BUSI in H_2O (0.016 M, pH 4.9, 45°C; 500 MHz; absolute value) (from Wider et al., 1984).

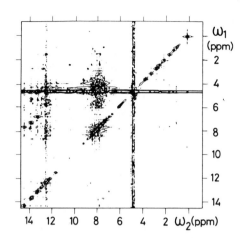

Figure 5.38. Solvent suppression in the ^{1}H NOESY spectrum of yeast tRNA^{Phe} in H_2O by combined use of a semiselective observation pulse (Fig. 5.2C) and digital shift accumulation (0.002 M, 0.005 M MgCl$_2$, 0.01 M Na$_2$HPO$_4$, 0.0001 M EDTA, 0.08 M NaCl, pH 6.9, 28°C; 500 MHz; absolute value; total acquisition time 66 h) (from Hilbers et al., 1983b).

factor of 200 to 500 (Haasnoot and Hilbers, 1983). Further suppression of the water peak by a factor of 50 to 200 can then be achieved by digital shift accumulation (Roth et al., 1980), so that overall a suppression by a factor of 1×10^4 to 1×10^5 can be obtained. Figure 5.38 shows a ^{1}H NOESY spectrum recorded with this combined procedure in a H_2O solution of tRNA. Similar to the spectra obtained with water saturation (Fig. 5.37), this spectrum contains a strong band of t_1 noise at the position of the H_2O resonance. Furthermore, as a consequence of the experimental procedure the spectrum is intrinsically asymmetric so that, for example, symmetrization would lead to artifacts (Haasnoot and Hilbers, 1983). Particular care must be exercised in the spectral analysis, since in addition to NOE cross peaks the spectra may contain cross peaks manifesting chemical exchange of labile protons with the bulk water (Haasnoot and Hilbers, 1983; Schwartz and Cutnell, 1983; see also Chapter 6). Alternate schemes for suppression of the H_2O resonance use the *jump-and-return* technique (Plateau and Guéron, 1982), and the potentialities of subspectrum selection by going through excitation of higher-order coherence (e.g., Basus, 1984; Guittet et al., 1984; Prestegard and Scarsdale, 1985).

5.8. HETERONUCLEAR 2D NMR

A wide range of 2D NMR experiments are available for studies of scalar and dipolar connectivities between different nuclei. Examples of applications to biological macromolecules include studies of ^{1}H–^{13}C scalar connectivities in proteins (Kojiro and Markley, 1983) and peptides (Kessler et al., 1983), ^{1}H–^{15}N heteronuclear COSY of nucleic acids (Griffey et al., 1983) and proteins (Live et al., 1984), sequential assignments in oligonucleotides by ^{1}H–^{31}P COSY (Marion and Lancelot, 1984; Pardi et al., 1983a), and studies of metal binding sites in proteins by ^{113}Cd–^{1}H COSY (Frey et al., 1985b; Live et al., 1985; Otvos et al., 1985). Overall, there is an almost unlimited range of potential applications for heteronuclear 2D NMR. In addition to adaptation of the principles followed in homonuclear 2D NMR, further aspects are of imminent interest, for example, enhancement of the sensitivity for observation of nonhydrogen nuclei by isotope enrichment and/or indirect detection (Bax et al., 1983; Bendall et al., 1983). However, since heteronuclear experiments did not so far have a major role for the main subject of this book, any attempt at a comprehensive coverage of heteronuclear 2D NMR would be beyond its scope. Instead, the following description of a particular experiment shall serve as an indication of the type of procedures used.

The metal binding in metallothionein-2 was investigated by studies of ^{113}Cd–^{1}H scalar couplings in the ^{113}Cd-enriched protein. Rabbit liver metallothionein-2 contains 20 Cys among its 62 amino acid residues, which are the coordination sites for seven ^{113}Cd^{2+} ions. The experimental scheme used (Fig. 5.39) is a modification of an established pulse sequence (Bax et al., 1983; Bendall et al., 1983; Bodenhausen and Ruben, 1980; Müller, 1979), which has also been employed in most other recent biological applications of heteronuclear 2D NMR. The experiment starts by creation of proton magnetization with a 90° pulse. After this initial coherence has evolved during the delay τ to antiphase magnetization with respect to ^{113}Cd, it is then transformed by a 90° ^{113}Cd pulse into heteronuclear two-spin coherence, which precesses during the evolution period t_1. The 180° ^{1}H pulse in the middle of t_1 refocuses the heteronuclear J interactions and ensures that only

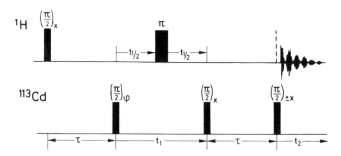

Figure 5.39. Scheme for ^{113}Cd–^{1}H COSY (note that π = 180°). The broken line in the ^{1}H channel represents a z filter. The phase φ is alternated to select ± 1Q coherence of ^{113}Cd, with alternated addition and subtraction of the free induction decays. This basic scheme can be modified for relayed coherence transfer (see text) or combined with other routines (e.g., TPPI, Fig. 5.19). [Reprinted with permission from Frey et al., *J. Am. Chem. Soc.*, **107**, 6847; © 1985 American Chemical Society.]

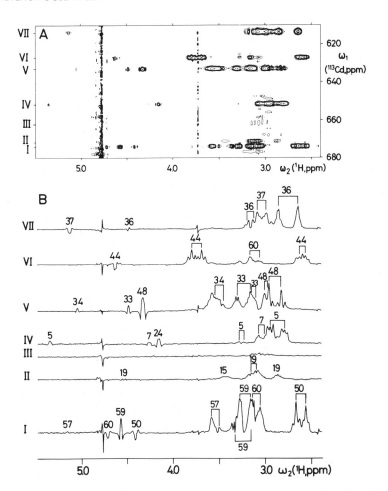

Figure 5.40. ^{1}H–^{113}Cd COSY spectrum of metallothionein-2 recorded with a z filter at the position of the broken line in Figure 5.39 (0.01 M, D_2O, 0.02 M [D_{11}]-Tris/HCl, 0.02 M KCl, pD 7.0, 20°C; 80 MHz ^{113}Cd, 360 MHz ^{1}H; τ = 8 ms, τ_z = 2 ms; phase sensitive). (A) Contour plot. The locations of the seven ^{113}Cd resonances are indicated on the left by the Roman numbers I–VII (B) Cross sections along ω_2 at the ω_1 positions of the seven ^{113}Cd resonances. βH resonances of Cys are between 2.5 and 4.0 ppm, αH resonances between 4.0 and 5.5 ppm. The numbers indicate the sequence locations of the cysteines. The resonance splittings due to ^{1}H–^{113}Cd coupling are indicated with solid brackets. Further splittings of the signals are due to ^{1}H–^{1}H scalar coupling. The spurious signals at 4.78 and 3.74 ppm are due to the residual protons in the solvent and the tris buffer. (Reprinted with permission from Frey et al., *J. Am. Chem. Soc.*, **107,** 6847; © 1985 American Chemical Society.)

the ^{113}Cd chemical shifts and the homonuclear couplings lead to modulations as a function of t_1. At the end of the evolution period a second 90° ^{113}Cd-pulse transfers the heteronuclear two-spin coherence back to proton coherence, which is observed after it was allowed to refocus during the second τ delay. For the experiments with metallothionein-2 this basic scheme was supplemented by two purging steps. A first purging process is caused by the insertion of a phase alternated 90° ^{113}Cd pulse prior to acquisition. In addition a z filter, $(90_y^\circ - \tau_z - 90_x^\circ)$, was inserted at the position of the broken line in Figure 5.39. Figure 5.40 shows a contour plot and cross sections along ω_2 through the seven ^{113}Cd ω_1 positions of a ^{113}Cd–^{1}H COSY spectrum of metallothionein-2. In addition to the improved sensitivity, the use of ^{1}H observation (Fig. 5.39) ensures that the superior digital resolution along ω_2 is used to unravel the complex ^{1}H spectrum. In the cross sections the individual cross peaks are attributed to connectivities between specified ^{113}Cd^{2+} ions and Cys residues. In addition to the direct connectivities between ^{113}Cd and βCH$_2$ of Cys, numerous weak relay peaks manifesting coherence transfer from ^{113}Cd to αCH are also present, which is a consequence of incomplete τ_z averaging in the z filter. If in the place of a z filter a 90_y° pulse is inserted at the position of the broken line (Fig. 5.39), the relay peaks will be favored relative to the direct connectivities (Frey et al., 1985b).

In a different approach, information on heteronuclear scalar coupling connectivities can be obtained by using the X nuclei as a relay station in homonuclear experiments. For example, ^{31}P-relayed ^{1}H COSY (Delsuc et al., 1984; Neuhaus et al., 1984b) was used for establishing through-bond connectivities between sequentially neighboring deoxyribose rings in oligodeoxynucleotides via the intervening ^{31}P nucleus (Frey et al., 1985a).

CHAPTER 6
Nuclear Overhauser Enhancement (NOE) in Biopolymers

The NMR approach to studies of biological macromolecules presented in this book is largely oriented around the ability of nuclear Overhauser enhancement (NOE) experiments to provide data on internuclear distances. These can be more directly correlated with the molecular conformation (Fig. 1.1) or with the structure of multimolecular complexes (Fig. 1.2) than other spectroscopic parameters. For the experimental observation of NOE's, three schemes were introduced in Figures 5.1B and C, and 5.4B. In *truncated driven* NOE (TOE) experiments (Fig. 5.1B) a resonance line A is subjected to weak, selective irradiation during a time period τ, which causes saturation of this resonance. This *preirradiation* is followed immediately by a nonselective observation pulse and acquisition of the free induction decay (FID). The NOE's resulting from the preirradiation on resonance A are manifested as fractional changes of the resonance intensities for nuclei located near nucleus A. Since these effects are usually difficult to detect in the complex spectra of biopolymers, a reference spectrum is recorded with preirradiation in an empty spectral region, and the *difference spectrum* containing only the lines with NOE's is then analyzed. In Figure 6.1 the preirradiation was applied to the well separated resonance of the aromatic ring protons 3,5H of Tyr-23 at 6.3 ppm in the ^{1}H NMR spectrum of BPTI. The TOE difference spectrum contains the αH and βCH$_3$ lines of Ala-25 and the resonances of two aromatic rings, which are all near the Tyr-23 ring in the spatial structure of BPTI. For obtaining the spectra of Figure 6.2, the 90° pulse in the scheme of Figure 5.1B was replaced by a semiselective observation pulse (Fig. 5.2) for suppression of the H$_2$O line. NOE's with hydrogen-bonded imino protons of base pairs in tRNA^{Phe} could thus be observed. In transient

93

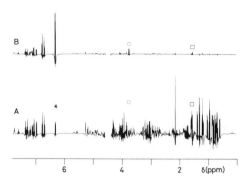

Figure 6.1. NOE's in BPTI (0.01 M, D_2O, pD 7.0, 35°C; 360 MHz; sine bell resolution enhancement). *(A)* Normal 1D ^1H NMR spectrum. *(B)* TOE difference spectrum ($\tau =$ 0.5 s). A star, circle, and square identify, respectively, the preirradiated line of Tyr-23, and the αH and βCH_3 lines of Ala-25. [Reprinted with permission from Wüthrich et al., *Biochemistry*, **17**, 2253-2263; © 1978 American Chemical Society.]

NOE experiments (Fig. 5.1*C*) the long preirradiation used for TOE measurements is replaced by a *selective 180° pulse* and a *delay interval* τ, and NOE difference spectra similar to those in Figures 6.1 and 6.2 can thus be obtained (Gordon and Wüthrich, 1978). In two-dimensional NOE (NOESY) experiments (Fig. 5.4*B*) the NOE's are manifested by cross peaks in the (ω_1, ω_2) plane (Fig. 5.37).

The NOE is a consequence of modulation of the dipole–dipole coupling

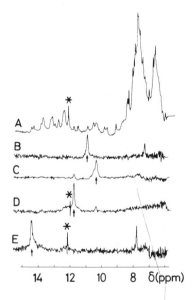

Figure 6.2. NOE's with labile protons in yeast tRNA^{Phe} (H_2O, 0.1 M NaCl, 0.01 M EDTA, 0.01 M cacodylate, pH 7.0, 21°C; 270 MHz). *(A)* Normal 1D ^1H NMR spectrum. *(B)*–*(E)* TOE difference spectra ($\tau = 0.3$ or 0.4 s). The preirradiated lines are identified with arrows. The star identifies an artifact (from Johnston and Redfield, 1978).

between different nuclear spins by the Brownian motion of the molecules in solution, and the NOE intensity can be related to the distance r between preirradiated and observed spin by an equation of the general form

$$\text{NOE} \propto \left\langle \frac{1}{r^6} \right\rangle \cdot f(\tau_c) \tag{6.1}$$

The second term in Eq. (6.1) is a function of the *correlation time* τ_c, which accounts for the influence of the motional averaging process on the observed NOE. Equation (6.1) would seem to indicate that distance measurements with the use of NOE's should be straightforward, provided that $f(\tau_c)$ can be independently assessed. In reality, a number of fundamental and technical obstacles tend to render quantitative distance measurements difficult. Thus, in all NOE experiments, and in particular in NOESY, processes other than NOE's may also be manifested and can lead to falsification of apparent NOE intensities. Quite generally, because of the low sensitivity for observation of NOE's, the accuracy of integration of line intensities is also limited by low S/N. Fundamental difficulties can then also arise when trying to correlate experimental NOE intensities with distances, for example, because of spin diffusion or the prevalence of intramolecular mobility in macromolecules. In the following these different aspects are considered in more detail.

6.1. NUCLEAR OVERHAUSER EFFECTS IN SMALL AND LARGE MOLECULES

The NOE phenomenon is intimately related to spin relaxation (Abragam, 1961; Noggle and Schirmer, 1971; Solomon, 1955). Analogous to the spin relaxation times T_1 and T_2 (Fig. 3.1), the NOE varies as a function of the product of the Larmor frequency ω_0, and the rotational correlation time τ_c. In Figure 6.3 we consider the situation of a pair of closely spaced spins i and j, connected by the vector $\mathbf{r}_{ij}$, and located either in a small or large spherical molecule. As a result of the collisions with the surrounding solvent and solute molecules, the thermal motions of these spheres consist of a random walk, which includes both translational and rotational movements. The relevant quantity for dipole–dipole relaxation and NOE is the rotational tumbling of the vector $\mathbf{r}_{ij}$, and the concomitant time variation of the angle θ_{ij} between $\mathbf{r}_{ij}$ and $\mathbf{B}_0$. If the mobility of this vector is restricted to the overall rotations of the molecule, $\mathbf{r}_{ij}$ will change orientation much more frequently in the small molecule than in the large molecule. For spherical particles of radius a in a solvent of viscosity η, a correlation time characterizing the frequency range for these stochastic motions can be estimated as

$$\tau_c = \frac{4\pi\eta a^3}{3kT} \tag{6.2}$$

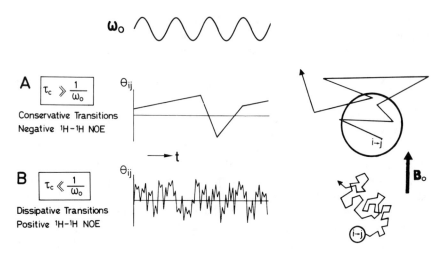

Figure 6.3. Brownian motion and NOE. The scheme considers two spins i and j, separated by the vector $\mathbf{r}_{ij}$ and located either in a large *(A)* or small *(B)* spherical molecule. θ_{ij} is the angle between $\mathbf{r}_{ij}$ and the magnetic field $\mathbf{B}_0$. The drawings on the right are a pictorial presentation of the rotational motions of $\mathbf{r}_{ij}$, where the length of the segments in the random walk indicates the time after which a distinct rotation state is again interrupted by collisions with other molecules. In the center, the variations of θ_{ij} are plotted versus time, and compared with the Larmor frequency of the spins ω_0.

In small molecules, for example, a tripeptide or a dinucleotide in aqueous solution, τ_c is short relative to ω_0^{-1} (at 500 MHz, $\omega^{-1} = 3 \times 10^{-10}$s). In this *extreme motional narrowing* situation, the frequency range covered by the rotational motions of $\mathbf{r}_{ij}$ includes ω_0 and $2\omega_0$ (Fig. 6.3*B*), which enables dissipative transitions between different spin states. In contrast, for macromolecules τ_c is long relative to ω_0^{-1} (Fig. 6.3*A*), and the frequencies of the rotational motions are too low to allow efficient coupling with the nuclear spin transitions. Therefore, energy-conserving transitions of the type $\alpha_i\beta_j \rightarrow \beta_i\alpha_j$ *(cross relaxation)* are favored.

The four curves in Figure 6.4 describe the maximum NOE's for the nuclear spins in Table 2.1 interacting with ^{1}H, if the preirradiation is on ^{1}H and the relaxation of the observed spin is entirely by dipole–dipole coupling with the preirradiated proton. For extreme motional narrowing, this limiting NOE factor for spin i with preirradiation of spin j is

$$n_i\{j\} = \frac{\gamma_j}{2\gamma_i} \tag{6.3}$$

In 1D experiments with relative line intensities 1 in the absence of NOE's, the line intensities *with* NOE then become

$$I_i = 1 + n_i\{j\} \tag{6.4}$$

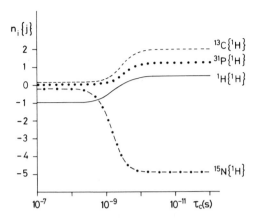

Figure 6.4. Plots of the maximum NOE (Eq. 6.3) versus log τ_c for ^1H, ^{13}C, ^{15}N, and ^{31}P interacting with ^1H. $\mathbf{B_0} = 11.74$ T, preirradiation on ^1H, it is assumed that the relaxation is entirely by dipole–dipole coupling with the preirradiated proton.

The $^1H\{^1H\}$ NOE, which is of prime interest for conformational studies, is $+0.5$ for the extreme motional narrowing situation. For τ_c longer than approximately 1×10^{-9} s it adopts a value of -1.0. (For lower fields B_0, the transition between these limiting values will occur at somewhat longer τ_c than in Fig. 6.4.) For ^{13}C and ^{31}P, $n_i\{^1H\}$ is positive over the entire τ_c range and becomes very small for long τ_c. For ^{15}N, $n_i\{^1H\}$ is negative throughout because of the negative value of γ (Table 2.1). The NOE's $^1H\{^{31}P\}$, $^1H\{^{13}C\}$, and $^1H\{^{15}N\}$, are very small [Eq. (6.3)] and of little practical importance in macromolecules.

In general, mechanisms other than dipole–dipole coupling with the preirradiated spin contribute to the T_1 relaxation (Abragam, 1961). If $T_{1d}(j)$ accounts for the dipolar relaxation between spins i and j and T_{1o} for all other contributions to T_1 of spin i, Eq. (6.3) becomes

$$n_i\{j\} = \frac{\gamma_j}{2\gamma_i} \frac{T_{1d}(j)^{-1}}{T_{1d}(j)^{-1} + T_{1o}^{-1}} \qquad (6.5)$$

Accordingly, the NOE can be partially or completely *quenched* in the presence of alternate, efficient relaxation pathways, for example, through proximity of spin i to a paramagnetic center. Again, Eq. (6.5) describes the maximum NOE that could be observed in the presence of the relaxation pathways T_{1o}.

In a network of like spins contained in a macromolecule with $\tau_c \gg \omega_0^{-1}$, *spin diffusion* by two or several subsequent cross relaxation steps can greatly influence the observed NOE intensities (Hull and Sykes, 1975; Kalk and Berendsen, 1976; Solomon, 1955). In the simple example of three spins shown in Figure 6.5 a two-step pathway for cross relaxation, spin 1 to spin 2

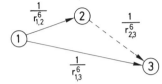

Figure 6.5. Direct cross relaxation between two spins 1 and 2, and 1 and 3 (solid arrows) and spin diffusion pathway from spin 1 via spin 2 to spin 3 (broken arrow).

followed by spin 2 to spin 3, may under certain experimental conditions be more efficient than direct cross relaxation between spins 1 and 3. The NOE on spin 3 is then no longer a faithful manifestation of the internuclear distance $r_{1,3}$. In the spatial structures of proteins and nucleic acids, the geometric arrangement of hydrogen atoms usually allows for a variety of spin diffusion pathways in addition to direct cross relaxation between distinct groups of protons.

6.2. NOE BUILDUP AND SPIN DIFFUSION

Incoherent exchange of magnetization by cross relaxation is a slow process, and the buildup of the NOE's can typically be followed on a time scale of seconds. A first illustration is provided by TOE experiments with BPTI (Fig. 6.6A). The preirradiation was on the well-separated amide-proton line of Tyr-23 at the low field end of the spectrum. With $\tau = 50$ ms, only a single NOE can be unambiguously detected. This line, which is identified by a filled circle, grows steadily with increasing length of the preirradiation period, and additional lines also appear. The τ dependence for the NOE's on three α protons is plotted in Figure 6.6B. As a second example Figure 6.7A shows plots of the τ dependence of transient NOE's in horse ferrocytochrome c. The ^{1}H NMR spectrum of this protein (Fig. 3.6) contains several well separated lines between -4 and -1 ppm, which correspond to the ring current shifted resonances of a methionine bound to the heme iron (Fig. 6.7B). The 180° pulse (Fig. 5.1C) was applied to the ε-methyl resonance near -3.2 ppm (Fig. 3.6). The curves for the individual β and γ protons show a rapid initial NOE buildup, which soon levels off, so that the NOE passes through a maximum and then decays. A further important observation is that the buildup curve for the α proton extrapolates to a starting point at a τ value of approximately 60 ms, rather than to $\tau = 0$ ms.

The crucial difference between the two experiments is that no radio-frequency (rf) field is applied during the NOE buildup in the transient NOE scheme, whereas the buildup is *driven* by continuous irradiation in the TOE scheme (Fig. 5.1B and C). Therefore, the decay of the NOE's by spin relaxation is readily observable in the transient NOE experiment, where no further energy is transferred to the nuclear spins after the initial 180° pulse. In contrast, a *steady state* between NOE buildup by the continuous irradiation

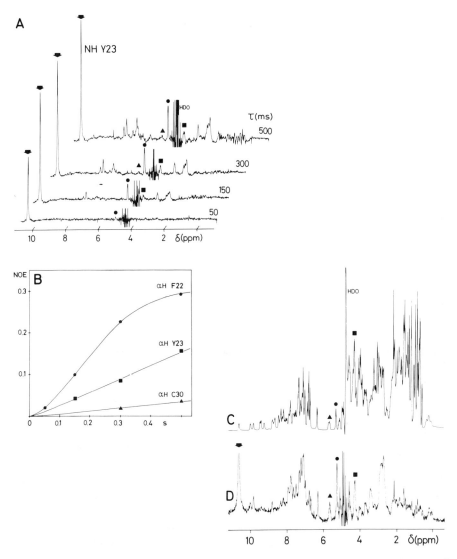

Figure 6.6. NOE buildup observed in 1D TOE difference spectra of a freshly prepared D₂O solution of BPTI (0.02 *M*, pD 4.5, 15°C; 360 MHz). *(A)* TOE difference spectra recorded with preirradiation on the lowest-field amide-proton line (arrow) during different intervals, τ. *(B)* Plots of NOE intensity versus τ for the three lines identified in *(A)*. *(C)* Normal 1D ¹H NMR spectrum. *(D)* Steady-state NOE difference spectrum obtained with selective preirradiation (arrow) during 4 s (from Dubs et al., 1979).

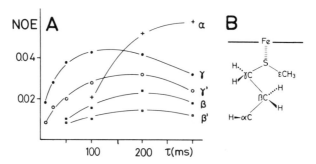

Figure 6.7. NOE buildup observed in 1D transient NOE difference spectra of horse ferrocytochrome c. *(A)* Plots of NOE intensity versus τ for the α, β, and γ protons of the axial methionine after applying a 180° pulse to the ε-methyl resonance. *(B)* Structure of the axial Met.

and NOE decay by spin relaxation is attained for long preirradiation times in the TOE scheme. [In the place of TOE, the term *presteady-state NOE* is therefore also in use (e.g., Clore and Gronenborn, 1985b).]

Steady-state NOE experiments using the scheme of Figure 5.1B with $\tau \gtrsim 5T_1$ have long been used for studies of small molecules (Noggle and Schirmer, 1971) and were also applied for detailed studies of oligopeptides (e.g., Glickson et al., 1976; Kuo and Gibbons, 1980; Leach et al., 1977). They have the advantage of optimal sensitivity for the observation of NOE's according to Eq. (6.5). For work with macromolecules in the regime of *negative* ^{1}H–^{1}H NOE's (Fig. 6.4), this advantage cannot be exploited because of spin diffusion. This is strikingly demonstrated by the steady-state NOE difference spectrum of BPTI in Figure 6.6D. In addition to the lines that were also observed when the rf-driven NOE buildup was truncated after short preirradiation times τ (Fig. 6.6A), steady-state NOE's prevail for a large percentage of all resonance lines in the normal ^{1}H NMR spectrum (Fig. 6.6C). Obviously, the buildup of most steady-state NOE's must have started only after the preirradiation had been applied during several hundred milliseconds. This can readily be rationalized by further inspection of Figure 6.5. In the spin diffusion pathway, magnetization is first transferred from the preirradiated spin 1 to spin 2. The second transfer from spin 2 to spin 3 will only be effective after buildup of a sizable NOE on spin 2, and therefore it will be delayed. Similarly, the delay in the transient NOE build-up curve for αH of Met in ferrocytochrome c (Fig. 6.7) shows that the NOE between the methionine ε-methyl group and the α proton is largely due to spin diffusion.

While in steady-state NOE's of macromolecules the distance information is usually masked by spin diffusion, theory shows that the initial NOE buildup rate is directly related to the distance between preirradiated and observed spin (Noggle and Schirmer, 1971; Solomon, 1955; Wagner and

Wüthrich, 1979). For like spins the time dependence of the magnetization on the nonirradiated spin 2 is

$$\frac{dM_2}{dt} = - \rho_2 M_2 - \sum_{j \neq 2} \sigma_{2j} M_j \qquad (6.6)$$

M_2 is the difference between the actual magnetization M_2^z of spin 2 and its equilibrium magnetization M_2^0, and M_j represents the same quantity for the spins j. The quantities ρ_2 and σ_{2j} determine the spin-lattice relaxation and the cross relaxation, respectively,

$$\rho_2 = \frac{\hbar^2 \gamma^4}{10} \sum_{j \neq 2} \frac{1}{r_{2j}^6} \left[\tau_c + \frac{3\tau_c}{1 + (\omega_0 \tau_c)^2} + \frac{6\tau_c}{1 + 4(\omega_0 \tau_c)^2} \right] \qquad (6.7)$$

and

$$\sigma_{2j} = \frac{\hbar^2 \gamma^4}{10} \frac{1}{r_{2j}^6} \left[\frac{6\tau_c}{1 + 4(\omega_0 \tau_c)^2} - \tau_c \right] \qquad (6.8)$$

The summation in Eq. (6.7) is over all spins j in the molecule considered. For $t = 0$, we have in Eq. (6.6) $M_2 = 0$ and all but one $M_j = 0$, the exception being the magnetization on the preirradiated spin M_1. Equation (6.6) then becomes

$$\left. \frac{dM_2}{dt} \right|_{t=0} = - \sigma_{21} M_1 \qquad (6.9)$$

and the initial slope of the NOE buildup curve is thus equal to the cross relaxation rate σ_{21}. In practice there is usually a short induction period in the TOE build-up curves, because the saturation of spin 1 is not instantaneous (Fig. 6.6B) (Wagner and Wüthrich, 1979). Furthermore, for very short τ values the S/N tends to be too low for quantitative measurements of line intensities. Therefore the initial slope is approximated by extrapolation of the data collected with τ up to ~0.3 s in TOE difference spectra (Fig. 6.6A and B), or with τ up to ~0.05 s in 1D transient NOE difference spectra (Fig. 6.7).

TOE difference spectra have been widely used for studies of biopolymers, and numerous NOE buildup curves in proteins and nucleic acids were measured with this technique (e.g., Dubs et al., 1979; Olejniczak et al., 1984; Scheek et al., 1984; Senn et al., 1984) and compared with buildup curves computed for particular geometric arrangements of groups of spins (e.g., Bothner-By and Noggle, 1979; Clore and Gronenborn, 1985c; Dobson et al., 1982; Wagner and Wüthrich, 1979). The technique is still valuable for work with well separated lines, such as the high-field region of the ferrocytochrome c spectrum (Fig. 3.6). In typical biopolymer spectra, however, TOE data may be falsified by limited selectivity of the preirradiation. Therefore, NOESY will in most instances be the preferred technique. Finally, it should be added that the two experiments of Figure 5.1*B* and *C* can also be used to measure magnetization transfer by chemical exchange (Redfield and Gupta, 1971) and that care must be exercised for proper distinction of transfer by exchange or by the NOE. [In certain situations, unambiguous distinction between exchange and cross relaxation can be obtained by 1D *zz* exchange spectroscopy (Wagner et al., 1985a).]

6.3.   TWO-DIMENSIONAL NUCLEAR OVERHAUSER ENHANCEMENT AND EXCHANGE SPECTROSCOPY (NOESY)

The initial steps in the presentation and analysis of ^{1}H NOESY spectra are similar to those used for COSY (Sections 5.2 and 5.3). Again, combined use of contour plots and cross sections is more suitable for the spectral analysis than the stacked plot presentation (Fig. 5.37). Similar to COSY, the NOESY cross peaks connect individual diagonal peaks. Numerous NOESY cross peaks are expected to occur in the same locations as COSY cross peaks, since hydrogen atoms connected by scalar coupling are usually also located at a short through-space distance. In addition, spectral regions that are nearly or completely empty in the COSY spectrum, may be crowded with NOESY cross peaks.

The protein ^{1}H NOESY spectrum of Figure 6.8 illustrates how the cross peaks are correlated with distinct diagonal peaks. All NOE's with a specified proton are manifested on the cross sections along ω_1 and ω_2 through the diagonal peak of this proton. Above the diagonal, connectivities to protons at lower field are on the cross section along ω_2 and those to protons at higher field on the cross section parallel to ω_1 (broken lines through the diagonal peak at 7.2 ppm in Fig. 6.8). The dotted lines show how the diagonal positions of individual coupling partners of the proton at 7.2 ppm are identified. Below the diagonal the spectrum is divided into six areas containing the following connectivities:

   a. NH; aromatics–NH; aromatics.
   b. NH; aromatics–αH; δH of Pro; βH of Ser and Thr
   c. NH; aromatics–aliphatic side chains
   d. αH; δH of Pro; βH of Ser and Thr–αH; δH of Pro; βH of Ser and Thr
   e. αH; δH of Pro; βH of Ser and Thr–aliphatic side chains
   f. Alphatic side chains–aliphatic side chains

To minimize interference with the solvent resonance, the regions d, e, and f are preferably studied in D_2O solution. After complete exchange of the amide protons, the regions a, b, and c are greatly simplified in a D_2O spectrum, since they then contain exclusively connectivities between aromatic protons and the other nonlabile hydrogens. Comparison of such a D_2O spectrum with the spectra recorded in H_2O affords an unambiguous identification

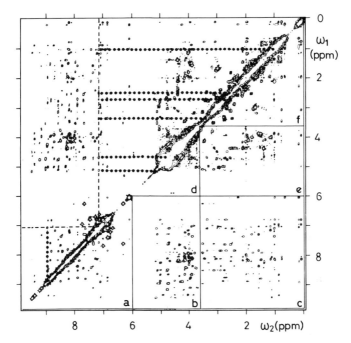

Figure 6.8. Protein ^{1}H NOESY spectrum (same as Fig. 5.37, symmetrized). Above the diagonal, broken and dotted lines indicate how connectivities between diagonal peaks are established via the cross peaks. Below the diagonal, the spectrum is divided into six regions a–f, which contain cross peaks between distinct proton types (see text).

of the cross peaks involving labile protons. Furthermore, comparison of corresponding NOESY and COSY spectra allows one to identify the NOE's between J-coupled hydrogen atoms, which is usually of great help in the initial stages of the spectral analysis.

The NOESY spectrum of a DNA duplex in D_2O solution (Fig. 6.9) was, on the basis of the chemical shift ranges for the different proton positions (Table 2.6), divided into 21 regions containing cross peaks between distinct proton types. For example, the region a contains cross peaks connecting 2H or 8H of a purine base or 6H of a pyrimidine base with methyl groups of T. The types of protons connected by the cross peaks in the other regions can be identified from the resonance assignments indicated at the top and on the left of the figure. In a spectrum of a DNA duplex recorded with a relatively long mixing time, nearly all of these spectral regions contain numerous cross peaks (Fig. 6.9). In ^{1}H NOESY spectra of nucleic acids recorded in H_2O (Fig. 5.38), additional cross peaks are expected in the spectral regions ($\omega_1 = 10–15$ ppm, $\omega_2 = 10–15$ ppm) for NOE's between different hydrogen-bonded imino protons, ($\omega_1 = 6.5–9.5$ ppm, $\omega_2 = 10–15$ ppm) for NOE's between imino and amino protons, ($\omega_1 = 6.5–9.5$ ppm, $\omega_2 = 6.5–9.5$ ppm) for NOE's

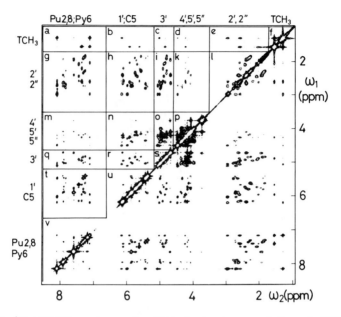

Figure 6.9. ¹H NOESY spectrum of a DNA duplex [same solution of d(CGCGAAT-TCGCG)₂ as in Fig. 5.12; 500 MHz, τ_m 300 ms; absolute value, symmetrized] (from Hare et al., 1983). Above the diagonal, the spectrum is divided into regions a–v, which contain cross peaks between distinct proton types, as indicated on the left and at the top.

between different amino protons and (ω_1 = 0–8.5 ppm, ω_2 = 6.5–15 ppm) for NOE's connecting labile with nonlabile protons.

In NOESY (Fig. 5.4B) the NOE's are built up during the mixing time τ_m, which separates the second and third 90° pulses. In analogy to the 1D transient NOE experiment, no rf irradiation is applied during the period between the mixing pulse and the observation pulse, and therefore similar buildup curves are obtained with these two experiments (Figs. 6.7 and 6.10). In NOESY the behavior of the diagonal peaks corresponds to that of the lines that are inverted by the 180° pulse in 1D transient NOE experiments (Fig. 5.1C). In Figure 6.10B the broken line shows the decay of the diagonal peak of Phe-33 NH, which is the result of cross relaxation and possibly other relaxation pathways. During the time interval of 300 ms, numerous cross peaks are built up. Those of Thr-32 αH and βH, Phe-33 2,6H, and Arg-20 NH result from direct cross relaxation with Phe-33 NH, whereas the delayed buildup for Thr-32 γCH₃ and Ile-19 γCH₃ shows that the observed NOE's are primarily the result of spin diffusion.

For work with biological macromolecules it is particularly attractive that a single NOESY experiment, which uses exclusively nonselective rf pulses (Fig. 5.4B), provides the complete network of NOE connectivities between

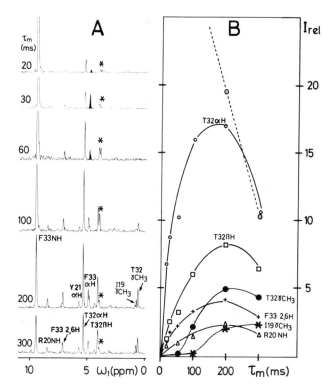

Figure 6.10. NOE buildup observed in NOESY spectra of BPTI (0.02 *M*, D$_2$O, pD 4.6, 24°C; 360 MHz). (*A*) Cross sections along ω_1 at ω_2 = 9.39 ppm (NH chemical shift for Phe-33) obtained with different mixing times. Distinct NOE peaks are identified with residue type, sequence position, and proton type. The star identifies the tail of a peak that is centered in an adjoining cross section. The black lines correspond to J peaks. (*B*) Plots of relative peak intensities in the spectra (*A*) versus τ_m. The broken line describes the decay of the diagonal peak of Phe-33 NH [Reprinted with permission from Anil-Kumar et al., *J. Am. Chem. Soc.*, **103**, 3654–3658; © 1981 American Chemical Society.]

distinct protons that prevail for a particular mixing time τ_m. Special precautions must be taken, however, to prevent loss or falsification of part of the information content of NOESY spectra due to certain additional, potentially detrimental features of the experiment, which are briefly discussed in the following.

1. Similar to the 1D NOE experiments, NOESY registers effects arising from incoherent transfer of magnetization by cross relaxation as well as by chemical or physical exchange processes. A typical example is the flipping motion of aromatic rings in proteins (Figs. 1.5 and 1.6). In the regime where separate resonance lines are observed for the individual 2,6- or 3,5- ring protons (Fig. 1.5), exchange cross peaks connecting the symmetry-related protons may be observed. [The experiment of Figure 5.4B is also often referred to as *2D exchange spectroscopy* (Jeener et al., 1979). With small molecules, applications for studies of chemical exchange may quite possibly turn out to be even more useful than NOE measurements (Macura and Ernst, 1980).] When studying NOE's in biopolymers it is therefore important that the potential occurrence of exchange processes other than by cross relaxation is recognized at the outset of an investigation. Often the different exchange pathways can then be distinguished based on the different dependence on experimental conditions, such as pH or temperature. In certain situations, *zz spectroscopy* allows one to unambiguously distinguish between NOE cross peaks and other exchange phenomena in complex molecular systems (see *3*). This distinction might also be achieved in rotating frame NOE experiments (Bothner-By et al., 1984; Davis and Bax, 1985).

2. In systems containing groups of scalar-coupled spins, NOESY spectra recorded with the pulse sequence of Figure 5.4B may contain additional cross peaks originating from coherent transfer of magnetization (Macura et al., 1981). The mechanisms that produce these unwanted *J peaks* are schematically illustrated in Figure 6.11. In addition to incoherent exchange, a sequence of two 90° pulses separated by an interval t_1 generates zero-quan-

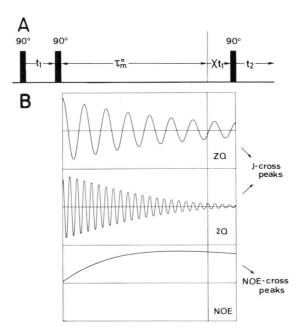

Figure 6.11. *(A)* Experimental scheme for NOESY modified for J-peak suppression by τ_m incrementation. Parallel to t_1 the mixing time is incremented by χt_1, where $(\chi t_1)_{max}$ is typically shorter than $\tau_m^o/10$. *(B)* Evolution of zero-quantum coherence, 2Q coherence, and NOE during τ_m (from Macura et al., 1982a).

tum (ZQ) and multiple-quantum (MQ) coherence. These coherence components evolve in an oscillatory fashion (Fig. 6.11B) and are then partially transformed into observable transverse magnetization by the third 90° pulse. An illustration is provided by the J peaks at 20, 30, and 60 ms in Figure 6.10A. Cross peaks that would result from MQ transitions can be suppressed by the proper choice of phase cycling (Macura et al., 1981; Wider et al., 1984), but separation of ZQ cross peaks from NOE's must rely on the different τ_m dependence of coherent and incoherent transfer processes (Fig. 6.11B). First, ZQ transfer will only be noticeable in NOESY spectra recorded with short mixing times (Macura et al., 1982b). Second, because of the oscillatory evolution of ZQ coherence, which contrasts with the slow, monotonous variation of the NOE with τ_m, the J cross peaks will be more extensively affected by small variations of τ_m during a NOESY experiment. For example, in Figure 6.11A, the mixing time is *incremented* parallel to t_1. Provided that χt_1 is short compared to τ_m^0, the resulting effect on the NOE's is hardly noticeable, whereas the J cross peaks are displaced parallel to ω_1. By subsequent symmetrization, the ZQ peaks can be removed from the NOESY spectrum (Macura et al., 1982a). Besides τ_m incrementation, random variation of τ_m during a NOESY experiment has also been used (Macura et al., 1981), and instead of varying the overall length of τ_m, a 180° pulse was inserted during τ_m (Macura et al., 1982b). Optimal J peak suppression can be attained by coaddition of signals from a series of measurements with such a refocusing pulse inserted at suitably chosen, different points in τ_m (Rance et al., 1985b).

3. When the rf pulses in a NOESY experiment deviate from 90°, longitudinal two-spin order, $I_{kz} I_{lz}$, may be created. In a conventional NOESY measurement, undesired extra peaks may thus be produced, which can be suppressed by accurate adjustment of the pulse lengths or by use of composite pulses (Bodenhausen et al., 1984b). Alternatively, a NOESY spectrum recorded with 45° pulses can be decomposed by suitable phase cycling into two subspectra containing, respectively, NOE-cross peaks and zz-cross peaks. In the complex nuclear spin systems of biopolymers, zz-exchange spectroscopy then allows one to differentiate unambiguously between exchange by chemical or physical processes, or by cross relaxation, provided a fragment with at least two scalar-coupled spins is transferred from one site to another in the exchange process (Wagner et al., 1985a).

4. Obviously the information obtained from NOESY depends on the extent to which individual cross peaks can be resolved and identified. In practice this may be limited by mutual overlap of different peaks in crowded spectral regions, or near the diagonal. The situation can be greatly improved with the use of phase-sensitive experiments instead of absolute value spectra. NOESY spectra can be obtained in the pure absorption mode, with the multiplet components for both diagonal peaks and cross peaks being in phase. Because of the intrinsically narrower linewidths, the resolution in absorption mode NOESY spectra is thus improved not only for closely spaced cross peaks but also for cross peaks near the diagonal (Fig. 6.12A and B; the squares identify corresponding cross-peak locations in the two spectra). Furthermore, the resolved multiplet structures of two or several resonance lines with nearly identical chemical shifts may be used for identification of cross peaks connecting distinct ones of these lines. The spectrum of the protein BUSI in Figure 6.12C was recorded with high digital resolution along ω_2. The fine structures of the nearly degenerate α-proton resonances of Ser-27, Ala-41, and Lys-56 are indicated at the top of the figure, and six cross peaks that were identified on the basis of the fine structure along ω_2 are labeled with the types and sequence positions of the connected protons. For situations where mutual overlap of resonance lines prevents unambiguous peak assignments even in high-resolution spectra, experiments combining incoherent magnetization transfer with coherent transfer steps have been proposed, for example, RELAYED-NOESY (Wagner, 1984) and 2Q-NOESY (van de Ven et al., 1985), as well as ω_1-decoupled NOESY (Sørensen et al., 1985).

5. The identification of cross peaks can also be limited by poor S/N. The sensitivity for observation of NOE's is intrinsically low and, in NOESY

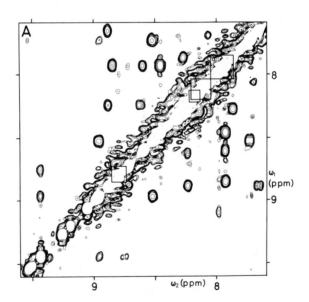

Figure 6.12. Improved spectral resolution in absorption-mode ^{1}H NOESY. *(A)* Low-field region of the absolute-value NOESY spectrum of BUSI in Figure 5.37. *(B)* Same region as in *(A)* from an absorption-mode NOESY spectrum of BUSI [same conditions, except that pH = 5.2, which caused several lines to shift relative to *(A)*]. *(C)* Expanded plot of a small region from a 500 MHz absorption-mode NOESY spectrum of BUSI recorded with a digital resolution of 0.4 Hz in ω_2 and 3.7 Hz in ω_1 (from Williamson et al., 1984, 1985).

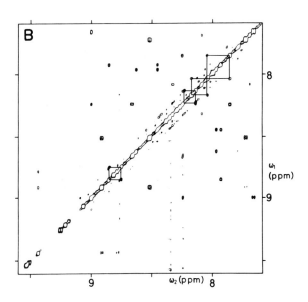

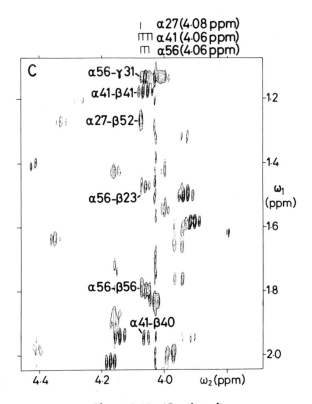

Figure 6.12. (Continued)

spectra recorded with short mixing times, t_1 ridges and t_2 ridges (Section 5.4.6) have up to now presented even more severe limitations than the thermal noise. As an illustration, Figure 6.13 shows a comparison of a NOESY spectrum of BPTI recorded using a mixing time of 40 ms with a corresponding, improved spectrum. In the improved spectrum the diagonal peaks and the t_1 ridges are suppressed as a result of a linear combination of three experiments recorded with different parameters (Denk et al., 1985). Clearly, numerous cross peaks can be identified in the improved spectrum, which are buried either under the diagonal or under the t_1 ridges in Figure 6.13A. As was described in Section 5.4.6, one can expect that t_1 ridges and t_2 ridges will in the future be more directly eliminated by proper performance of the discrete 2D FT (Otting et al., 1986), and alternative difference spectroscopy techniques for suppression of t_1 noise (Klevit, 1985) and diagonal peaks (Harbison et al., 1985) have also been proposed.

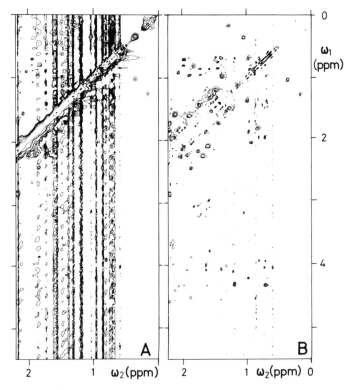

Figure 6.13. Combined suppression of diagonal peaks and t_1 ridges in NOESY. *(A)* Normal absorption-mode NOESY spectrum of BPTI (0.02 *M*, D_2O, pD 4.6, 36°C; 500 MHz; τ_m 40 ms;). *(B)* Spectrum recorded with diagonal-suppression (from Denk et al., 1985).

6.4. NUCLEAR OVERHAUSER EFFECTS, ¹H–¹H DISTANCES, AND MOLECULAR DYNAMICS

A variety of factors on different levels of data acquisition and analysis will eventually determine to which extent NOE's can be employed for quantitative measurements of ¹H–¹H distances. While important insights resulted from work done during the last few years, further quantitation of NOE distance measurements can be expected to be the subject of intensive studies in the near future. Therefore, the main purpose of this section is to define areas where further advances might soon be anticipated.

The initial step toward quantitative NOE experiments is the proper selection of the experimental schemes used, ensuring separation of direct NOE's from other processes that might contribute to the observed peak intensities. With the procedures of Section 6.3 this problem seems to be largely solved, except for special situations where differentiation of NOE's and chemical exchange processes might require additional controls. To account for cross-peak intensity arising from spin diffusion, analysis of NOESY build-up curves (Fig. 6.10) in the time range from approximately 20–200 ms appears to be the safest procedure. Incomplete suppression of J peaks and zz peaks exclusively affects distance measurements between scalar-coupled protons. These are of interest for the purpose of empirical calibration of NOE intensities versus distance and for studies of local conformation, but they hardly influence the determination of global features of biopolymer conformations.

A second step is the quantitative measurement of relative peak intensities. While evaluation of the integrated resonance intensity in 1D NMR is a standard procedure, NOESY peak intensities have so far mostly been estimated from measurements of the peak heights. Quantitation of relative cross peak intensities is then limited by factors such as poor S/N, different multiplet patterns depending on the interacting spins, overlap of two or several peaks, and poor digitization. Routines for peak deconvolution with suitable line shapes (Denk et al., 1986), or maximum-entropy processing (Laue et al., 1985; Hore, 1985) are in principle available for more quantitative spectral analyses, and alternative procedures have also been tried out (Broido et al., 1985). Considering the large data mass that needs to be processed, automated procedures for this step in the analysis are definitely needed.

Once quantitative NOE measurements at short mixing times are available, these can be expressed in terms of the corresponding cross relaxation rates between specified spins [Eq. (6.9)]. For two pairs of protons at distances r_a and r_b in a rigid structure, one would then have

$$\frac{\sigma_a}{\sigma_b} = \frac{r_b^6}{r_a^6} \tag{6.10}$$

Approximate validity of this relation was evidenced by numerous experiments with proteins for which the atomic coordinates in single crystals were shown to be largely preserved in solution. Subject to the limitations discussed on the following pages, Eq. (6.10) offers an avenue for determination of the distance ratios between different pairs of spins and thus for empirical calibration of the correlation between NOE and ^{1}H–^{1}H distance. An initial, straightforward application is for estimating an upper limit for NOE-observable distances in a rigid molecular structure. A useful standard for the maximum possible cross relaxation rate in biopolymers $\sigma(r_0)$ is the distance of 1.75 Å between two methylene protons, since this distance is less than the van der Waals contact distance between two nonbonded hydrogen atoms. We then define $Q(r_{ij}) = \sigma(r_{ij})/\sigma(r_0)$, where r_{ij} is the distance between two protons of interest, and plot $Q(r_{ij})$ versus r_{ij} (broken curve in Fig. 6.14). The smallest cross-relaxation rate $\sigma(r_{ij})$ that may be observed is determined by the sensitivity of NOE observation. In practice this is of the order of 5% of $\sigma(r_0)$. Figure 6.14 then indicates that in a perfectly rigid macromolecule, selective NOE's could be detected for pairs of protons that are at a distance of approximately 3.0 Å or less.

Equation (6.10) is valid only when identical *correlation functions*, $f(\tau_c)$, apply for $\mathbf{r}_a$ and $\mathbf{r}_b$. Internal motions in macromolecular structures that are too slow to affect $f(\tau_c)$ may then still be manifested in time variations of r_{ij}. An estimate of the influence of time fluctuations of the ^{1}H–^{1}H separation on the maximal NOE-observable distance can be obtained from a simple *uniform-averaging* model (Braun et al., 1981). This calculation assumes that the distances r_{ij} are uniformly spread over a range extending from the ^{1}H–^{1}H van der Waals distance of 2.0 Å to an upper limit R_m, with a corresponding cross relaxation rate $\sigma(R_m)$. A plot of $Q = \sigma(R_m)/\sigma(r_0)$ versus R_m (solid curve in Fig. 6.14) shows that NOE's corresponding to $R_m \lesssim 5.0$ Å should be observable. In reality, proteins and nucleic acids in solution at ambient temperature are neither rigid, nor will the steric constraints allow that the time fluctuations of ^{1}H–^{1}H distances extend all the way to the limit of 2.0 Å. Therefore, a more realistic upper limit for NOE-observable ^{1}H–^{1}H distances in biopolymers will lie somewhere in the area bounded by the two curves in Figure 6.14 and, depending on the system studied and the NMR equipment used, is of the order 3.5–4.5 Å. This distance limit is evidently short compared to the overall dimensions of proteins or nucleic acids, but on the other hand the

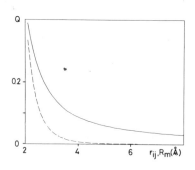

Figure 6.14. Dependence of the cross relaxation rate, represented by Q (see text), on the ^{1}H–^{1}H distance in a rigid molecular structure (broken curve) or in a flexible structure, assuming uniform averaging (solid curve) (from Braun et al., 1981).

restriction to short distances ensures that distinct peaks can be resolved in the NOESY spectra of macromolecules.

High frequency intramolecular motions can contribute to the overall rotational tumbling of the distance vectors $\mathbf{r}_{ij}$ and thus affect the effective correlation time. From the form of the correlation function for cross relaxation [Eq. (6.8)] it is apparent that for macromolecular systems, where the ^{1}H–^{1}H NOE is negative, the intramolecular angular changes can only reduce the magnitude of the NOE's (Braun et al., 1981; Olejniczak et al., 1984). Overall, intramolecular motions and alternative relaxation pathways always tend to reduce the NOE below the value that would correspond to the actual proton–proton distance and the correlation time for overall molecular rotations. Therefore, when working with a molecule with unknown conformation, weak NOE's, or the absence of NOESY cross peaks cannot a priori be used to establish lower limits on the spin–spin distances. Once a 3D structure is known with sufficient detail so that the environment of individual spins can be described, a complete description of all relaxation pathways can in principle be obtained. On this basis a more complete analysis of the observed NOE's, including partial or complete quenching of NOE's and effects due to spin diffusion, may then become warranted. This might primarily be of interest for refinement in advanced stages of a structure determination (Keepers and James, 1984; Broido et al., 1985).

Previous discussions on fundamental difficulties for the structural analysis that may arise from the nonlinear averaging of NOE's in flexible molecular structures were surveyed by Jardetzky and Roberts, 1981, pp. 139–140. The problem is probably difficult to solve for truly flexible, unstructured polypeptide or polynucleotide chains (where most NOE's can be expected to be quenched by intramolecular mobility). For molecules with defined conformations, it can be expected that theoretical studies using more complex models than the uniform-averaging calculation will in the future provide more reliable correlations between NOE's and molecular conformations. In particular, situations with rapid fluctuations between two or several distinct, preferentially populated conformers need to be analyzed in detail. The presently available experimental evidence indicates that more quantitative distance measurements will be needed for structure refinements, but that interpretation of ^{1}H–^{1}H NOE's in terms of qualitative upper constraints on intramolecular ^{1}H–^{1}H distances is sufficient to characterize the global folding of biopolymer chains (see Chapters 9, 10, 13, and 14).

PART II
RESONANCE ASSIGNMENTS AND STRUCTURE DETERMINATION IN PROTEINS

CHAPTER 7
NOE-Observable ^{1}H–^{1}H Distances in Proteins

This chapter presents a systematic study of short ^{1}H–^{1}H distances in poly-peptides, which provides a basis for conformational studies of proteins using NOE measurements (Chapters 8–10).

7.1. NOTATION FOR ^{1}H–^{1}H DISTANCES IN PROTEINS

The distance between the hydrogen atoms A and B located in the amino acid residues in the sequence positions i and j, respectively, is denoted by $d_{AB}(i,j)$. This notation can be applied for all hydrogen atoms (Table 2.2), but in practice the distances between different backbone hydrogen atoms, and between backbone hydrogens and βCH$_n$ groups are of particular interest.

$$d_{\alpha N}(i,j) \equiv d(\alpha H_i, NH_j) \tag{7.1}$$

$$d_{NN}(i,j) \equiv d(NH_i, NH_j) \equiv d_{NN}(j,i) \tag{7.2}$$

$$d_{\beta N}(i,j) \equiv \min\{d(\beta H_i, NH_j)\} \tag{7.3}$$

$$d_{\alpha \alpha}(i,j) \equiv d(\alpha H_i, \alpha H_j) \equiv d_{\alpha \alpha}(j,i) \tag{7.4}$$

$$d_{\alpha \beta}(i,j) \equiv \min\{d(\alpha H_i, \beta H_j)\} \tag{7.5}$$

The distances in Eqs. (7.1), (7.3), and (7.5) are defined in the direction from the N to the C terminus of the polypeptide chain (Fig. 7.1), and correspond-ing distances $d_{N\alpha}(i,j)$, $d_{N\beta}(i,j)$, and $d_{\beta\alpha}(i,j)$ will also be used. In the Eqs. (7.3) and (7.5), *min* indicates that in the case of β-methylene or β-methyl groups the shortest distance to any of the β protons is taken. [In publications up to

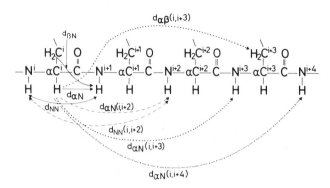

Figure 7.1. Selected sequential and medium-range ^{1}H–^{1}H distances in polypeptide chains (from Wüthrich et al., 1984a).

1984, $d_1(i,j)$, . . . , $d_5(i,j)$ were used instead of $d_{\alpha N}(i,j)$, . . . , $d_{\alpha\beta}(i,j)$; see Wüthrich et al. (1984a).]

The notation of Eqs. (7.1)–(7.5) provides an unambiguous description of all distances between different backbone protons, or between backbone protons and β protons in polypeptide chains consisting of common amino acid residues (Table 2.2), except for certain protons in glycine and proline. In the case of glycine α protons, the shorter distance is taken so that, for example, Eq. (7.1) becomes

$$d_{\alpha N}^{GX} (i,j) \equiv \min\{d(\alpha H_i, NH_j)\} \tag{7.6}$$

For proline, ^{1}H–^{1}H distances corresponding to those of Eqs. (7.1)–(7.5) can be defined with δCH_2 instead of the amide proton and taking the shorter one of the two distances to the δ-methylene protons. For example, Eq. (7.1) becomes

$$d_{\alpha\delta}^{XP} (i,j) \equiv \min\{d(\alpha H_i, \delta H_j)\} \tag{7.7}$$

Since all interresidue ^{1}H–^{1}H distances are across at least one peptide bond, their values may be different for cis or trans peptide bonds. Except for Figure 7.10, all data presented in this chapter are for trans peptide bonds.

The definitions of some interresidue ^{1}H–^{1}H distances in the primary polypeptide structure are illustrated in Figure 7.1. *Sequential* distances are those between backbone protons or between a backbone proton and a β proton in residues that are nearest neighbors in the amino acid sequence. For simplicity, the indices i and j are omitted for the sequential distances; for example, $d_{\alpha N}(i,i + 1) \equiv d_{\alpha N}$ and $d_{NN}(i,i + 1) \equiv d_{NN}$. *Medium-range* distances are all nonsequential interresidue distances between backbone protons or between a backbone proton and a β proton within a segment of five consecutive residues (Fig. 7.1). *Long-range backbone* distances are between backbone protons in residues that are at least six positions apart in the sequence, that

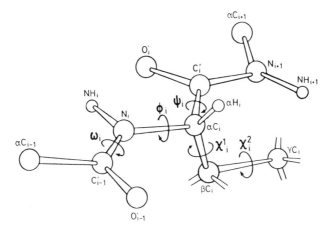

Figure 7.2. Standard nomenclature for the atoms and the torsion angles along a poly-peptide chain (IUPAC–IUB Commission on Biochemical Nomenclature, 1970).

is, $|j - i| \geq 5$. All other interresidue distances are referred to as *long-range* distances. This somewhat unorthodox classification is used because the me-dium-range and long-range backbone distances have a dominant role in stud-ies of polypeptide backbone conformations.

The following computations of ^{1}H–^{1}H distances use the ECEPP (empiri-cal conformational energy program for peptides) standard geometry for the individual amino acid residues (Momany et al., 1975; Némethy et al., 1983). Different polypeptide conformations are generated by variation of the tor-sion angles about single bonds (Fig. 7.2). In this procedure the polypeptide backbone is assumed to consist of planar trans peptide bonds linked together via the α-carbon atoms. All possible backbone conformations are obtained by variation of the torsion angles ϕ_i and ψ_i (Fig. 7.2; in addition there is in principle the possibility of cis–trans isomerization of peptide bonds, de-scribed by the torsion angles ω_i). The side chain conformations are charac-terized by the torsion angles $\chi_i^1, \chi_i^2, \ldots$.

7.2. INTRARESIDUE ^{1}H–^{1}H DISTANCES

Quite naturally, short ^{1}H–^{1}H contacts are ubiquitous within the individual amino acid residues. The distances between vicinal protons vary between approximately 2.15 and 2.90 Å, and the exact values are determined by the intervening torsion angle. For example, the distances $d_{N\alpha}(i,i)$ and $d_{\alpha\beta}(i,i)$ are related with ϕ_i and χ_i^1, respectively (Figs. 7.3 and 7.4). For $d_{N\alpha}(i,i)$ it is worth noting that similar values prevail for the major regular polypeptide second-ary structures. Two-dimensional plots of $d_{N\beta}(i,i)$ versus (ϕ_i, χ_i^1) and $d_{\alpha\gamma}(i,i)$ versus (χ_i^1, χ_i^2) were presented by Leach et al. (1977), and a complete listing

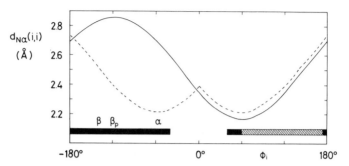

Figure 7.3. Intraresidue ¹H–¹H distance $d_{N\alpha}(i,i)$ versus the torsion angle ϕ_i. Solid curve, L-amino acid residues; broken curve, shorter NH–αH distance in Gly. The black bar indicates the sterically allowed ϕ_i values for L-amino acids; for Gly the region of the shaded bar is also allowed. α, β, and β_p indicate the ϕ_i values for the regular α helix, antiparallel β sheet, and parallel β sheet, respectively.

of the upper and lower bounds on the distances between any combination of two hydrogen atoms in the 20 common amino acid residues (Table 2.2) can be found in Wüthrich et al. (1983).

For obtaining ¹H NMR assignments, the distances between the β-methylene protons and the ring protons in Phe, Tyr, His, and Trp,

$$d_{\beta R}(i,i) \equiv \min\{d(\beta H_i, \text{Ring } H_i)\} \tag{7.8}$$

are of special interest, since these connectivities can in practice only rarely be established via the small, four-bond scalar couplings (Section 2.1, Table 2.2). Plots of $d_{\beta R}(i,i)$ versus χ_i^2 (Fig. 7.5) show that in Phe, Tyr, and Trp at least one ¹H–¹H distance between βCH_2 and the ring is shorter than 3.0 Å for all values of χ_i^2, whereas $d_{\beta R}(i,i)$ for His varies between 2.3 and 3.8 Å.

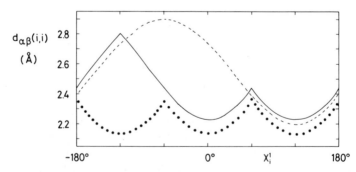

Figure 7.4. Intraresidue ¹H–¹H distance $d_{\alpha\beta}(i,i)$ versus the torsion angle χ_i^1. Solid curve, shorter αH–βH distance in αCH–βCH₂ fragments. Dotted curve, shortest αH–βH distance in Ala. Broken curve, Val, Ile, and Thr. [Because of the convention for numbering branched side chains (IUPAC–IUB Commission on Biochemical Nomenclature 1970), χ^1 in Val differs from χ^1 in Thr and Ile by +120° if βH is oriented identically relative to αH.]

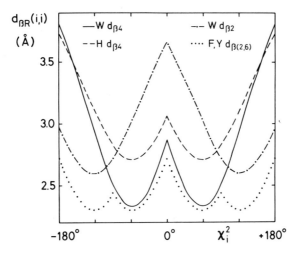

Figure 7.5. Intraresidue ^{1}H–^{1}H distances $d_{\beta R}(i,i)$ [Eq. (7.8)] in the aromatic residues Phe, Tyr, His, and Trp versus the torsion angle χ_i^2 (notation of the ring protons as in Table 2.2) (from Billeter et al., 1982).

With regard to resonance assignments, the distances between βCH_2 or γCH_2 and the side chain amide protons in Asn and Gln are also of interest. The geometry at the periphery of these side chains closely matches that of a backbone fragment with Gly, $-\alpha CH_2-CO-NH-$, and therefore plots of $d_{\beta\delta}(i,i)$ versus χ_i^2 or $d_{\gamma\varepsilon}(i,i)$ versus χ_i^3 for Asn or Gln, respectively, coincide almost quantitatively with the plot of $d_{\alpha N}^{GX}$ versus ψ_i (Fig. 7.6).

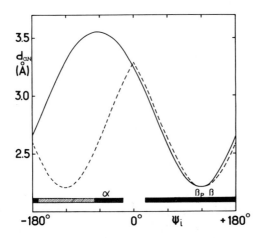

Figure 7.6. Sequential distance $d_{\alpha N}$ between residues i and $i + 1$ versus the torsion angle ψ_i. Solid curve, L-amino acid in position i. Broken curve, $d_{\alpha N}^{GX}$ [Eq. (7.6)]. For an explanation of the symbols near the bottom see Figure 7.3 (from Billeter et al., 1982).

7.3. SEQUENTIAL ^{1}H–^{1}H DISTANCES

The sequential distances $d_{\alpha N}$, d_{NN}, and $d_{\beta N}$ all depend on one or two intervening torsion angles about single bonds (Fig. 7.2). For L-amino acids, $d_{\alpha N}$
varies between approximately 2.20 and 3.55 Å for ψ_i values between -180
and $180°$, and the upper limiting value of $d_{\alpha N}^{GX}$ is near 3.30 Å (Fig. 7.6). For β-
pleated sheets, $d_{\alpha N}$ is close to the minimum value of 2.2 Å, for a regular
α helix it is approximately 3.5 Å.

Depending on the intervening backbone torsion angles ϕ_i and ψ_i, d_{NN}
varies between the van der Waals limit of 2.0 Å and approximately 4.7 Å
(Fig. 7.7). In the sterically allowed region A of the (ϕ_i,ψ_i) plane, d_{NN} is
shorter than 3.0 Å, with a value of 2.8 Å for the regular α helix. In region B,
d_{NN} varies between 2.0 and 4.8 Å, with values of 4.3 and 4.2 Å for the
antiparallel and parallel β structures, respectively. In region C, d_{NN} covers
the range 2.5–3.5 Å.

For $d_{\beta N}$ (Fig. 7.8), upper and lower limits are given, which were found by
variation of χ_i^1 from -180 to $180°$ for each value of ψ_i. It is seen that $d_{\beta N}$ can
adopt values shorter than 3.0 Å in the entire sterically allowed region A (Fig.
7.7). In region B it is between 2.2 and 4.5 Å. For a regular α helix $d_{\beta N}$ varies
between 2.5 and 4.1 Å, where for Ala the upper limit drops to approximately
3.0 Å (Fig. 7.8). In regular antiparallel and parallel β structures, $d_{\beta N}$ varies

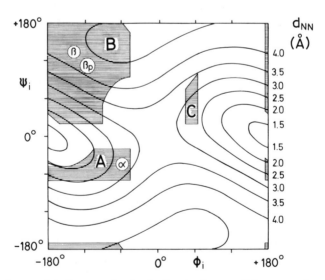

Figure 7.7. Sequential distance d_{NN} in the ϕ_i–ψ_i plane; solid contour lines represent
fixed values of d_{NN} as indicated on the right. The shaded areas A, B, and C are sterically
allowed for an alanyl dipeptide (Ramachandran and Sasisekharan, 1968). α, β, and β_p
indicate the ϕ_i–ψ_i combinations for the regular α helix, antiparallel β sheet, and parallel
β sheet (from Billeter et al., 1982).

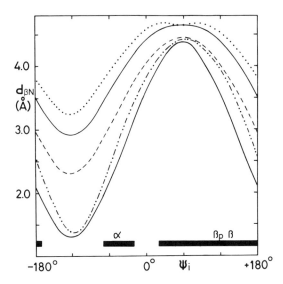

Figure 7.8. Sequential distance $d_{\beta N}$ versus the dihedral angle ψ_i. For each value of ψ_i, the range of distances $d_{\beta N}$ is given that is covered if χ_i^1 varies from -180 to $180°$. For L-amino acids with β-methylene groups, $d_{\beta N}$ varies between the two solid curves. For Val, Ile, and Thr, $d_{\beta N}$ is between the lower solid curve and the dotted curve. For a pseudoatom in the center of the three methyl protons of Ala, $d_{\beta N}$ lies on the broken line, and for a planar Pro ring it is on the line with dots and dashes. For an explanation of the symbols near the bottom see Figure 7.3 (from Billeter et al., 1982).

from 3.2 to 4.5 Å and 3.7 to 4.7 Å, respectively, with values of 3.6 and 4.0 Å for the shortest distance to Ala methyl protons.

Among the other sequential distances, $d_{\alpha\alpha}$ and $d_{N\alpha}$ are included in the following discussion on the influence of peptide cis–trans isomerization. $d_{N\beta}$, $d_{\alpha\beta}$, and $d_{\beta\alpha}$ cannot be shorter than 4.0 Å, except that $d_{\beta\alpha}$ covers the range 3.8–5.6 Å for the α helix ψ_i value of $-47°$.

Cis–trans isomerization of the intervening peptide bond has a dominant influence on the sequential ¹H–¹H distances. Since the occurrence of the cis form is most likely for the dipeptide segments Xxx-Pro and Pro-Pro, we consider these cases in more detail. However, the general, qualitative conclusions thus obtained would equally apply to any segment Xxx-Xxx with an intervening cis peptide bond.

Comparison of *cis* and *trans* Xxx-Pro (Fig. 7.9) shows that the distances between NH$_i$, αH$_i$, δCH$_{2(i+1)}$, and αH$_{(i+1)}$ are particularly strongly affected by the cis–trans isomerization. The cis form allows much closer contacts between αH$_i$ and αH$_{(i+1)}$, and between NH$_i$ and αH$_{(i+1)}$, whereas the trans form favors short distances between αH$_i$ and δCH$_{2(i+1)}$, and between NH$_i$ and δCH$_{2(i+1)}$. A quantitative presentation of these distances is given for Pro-Pro, since this enables one-dimensional plots for all four distances (Fig.

trans $\rightleftharpoons$ cis

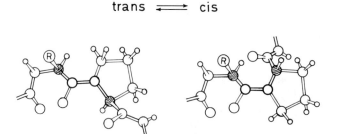

Figure 7.9. Cis–trans isomers of the Xxx–Pro peptide bond.

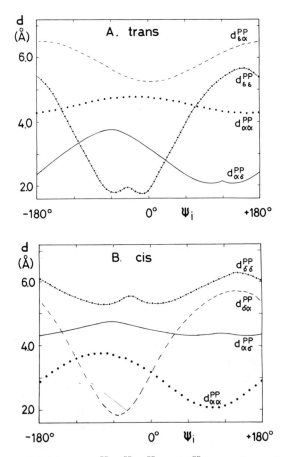

Figure 7.10. Sequential distances $d_{\alpha\delta}^{PP}$, $d_{\delta\alpha}^{PP}$, $d_{\alpha\alpha}^{PP}$, and $d_{\delta\delta}^{PP}$ versus the torsion angle ψ_i. (A) *trans* Pro-Pro. (B) *Cis* Pro-Pro (from Wüthrich et al., 1984a).

7.10). For *trans* Pro-Pro, the plot of $d_{\alpha\delta}^{PP}$ versus ψ_i is nearly identical to $d_{\alpha N}$ versus ψ_i in Figure 7.6. The plot for $d_{\delta\delta}^{PP}$ is similar to a corresponding plot for d_{NN}, but covers a somewhat larger distance range. (The corresponding plot of d_{NN} versus ψ_i would be the cross section parallel to the ψ_i axis through the contour lines in Fig. 7.7 at the ϕ_i position for a planar proline ring, i.e., 57.6°.) $d_{\alpha\alpha}^{PP}$ and $d_{\delta\alpha}^{PP}$ are longer than 4.2 Å and longer than 5.2 Å, respectively, over the entire ψ_i range. For *cis* Pro-Pro the corresponding plots (Fig. 7.10B) show that the behavior of $d_{\alpha\delta}^{PP}$ and $d_{\alpha\alpha}^{PP}$ is nearly quantitatively interchanged when compared to *trans* Pro-Pro, and the same applies for $d_{\delta\delta}^{PP}$ and $d_{\delta\alpha}^{PP}$.

7.4. SHORT ¹H–¹H DISTANCES IN REGULAR SECONDARY STRUCTURES

Inspection of common polypeptide secondary structures shows that in addition to the sequential distances, a variety of medium-range or long-range backbone ¹H–¹H distances are sufficiently short to be observed by NOE's. The α helix is characterized by close approach between residues i and $(i + 3)$, and between residues i and $(i + 4)$ (Fig. 7.11). In the 3_{10} helix, short distances prevail between residues i and $(i + 3)$, and between residues i and $(i + 2)$ (Fig. 7.11). In tight turns there are close contacts between residues i and $(i + 2)$ (Fig. 7.12). In β structures, the individual strands consist of almost fully extended polypeptide segments, which excludes short medium-range

$$\alpha(3.\,6_{13}) \qquad\qquad 3_{10}$$

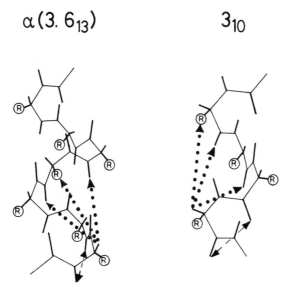

Figure 7.11 Short sequential and medium-range ¹H–¹H distances in the α helix and the 3_{10} helix. Broken arrow, d_{NN}. Dotted arrows, $d_{\alpha N}(i,i + 3)$, $d_{\alpha\beta}(i,i + 3)$, and $d_{\alpha N}(i,i +4)$ [α helix] or $d_{\alpha N}(i,i + 2)$ [3_{10} helix].

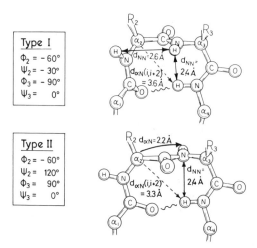

Figure 7.12. Short sequential and medium-range ¹H–¹H distances in type I and type II tight turns. The wavy lines indicate hydrogen bonds (from Wüthrich et al., 1984a).

distances, but the close approach of neighboring strands in β sheets brings about short interstrand ¹H–¹H distances (Fig. 7.13). Overall, helices and tight turns can be characterized by short sequential and medium-range ¹H–¹H distances, and β sheets by short sequential and long-range backbone ¹H–¹H distances. In helices and β sheets, the short ¹H–¹H contacts indicated in Figures 7.11 and 7.13 form a continuous, dense network over the entire length of these regular secondary structures.

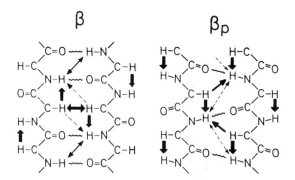

Figure 7.13. Short sequential and long-range backbone ¹H–¹H distances in β sheets. Wavy lines indicate interstrand hydrogen bonds. Thick vertical arrows indicate $d_{\alpha N}$. For antiparallel β, short interstrand distances are indicated by thick horizontal arrows $|d_{\alpha\alpha}(i,j)|$, thin solid arrows $|d_{NN}(i,j)|$, and broken arrows $|d_{\alpha N}(i,j)|$. In parallel β, solid arrows indicate $d_{\alpha N}(i,j)$ and broken arrows $d_{NN}(i,j)$ (from Wüthrich et al., 1984a).

In all, following the definitions introduced in Section 7.1, there are 8 different sequential ¹H–¹H distances and 24 different medium-range ¹H–¹H distances in polypeptide chains. A systematic study with standard secondary structures (Schulz and Schirmer, 1979; Richardson, 1981) generated from polyalanine with trans peptide bonds showed that 8 of these 32 distances (Fig. 7.1) adopt sufficiently short values to be observed by NOE's (Table 7.1) (Wüthrich et al., 1984a). In regular, nontwisted β sheets the long-range backbone distances indicated in Figure 7.13 adopt the following values:

Antiparallel β sheets, $d_{\alpha\alpha}(i,j) = 2.3$ Å

$d_{\alpha N}(i,j) = 3.2$ Å

$d_{NN}(i,j) = 3.3$ Å

Parallel β sheets $d_{\alpha N}(i,j) = 3.0$ Å

$d_{NN}(i,j) = 4.0$ Å

$d_{\alpha\alpha}(i,j) = 4.8$ Å

TABLE 7.1. Short ($\leq$ 4.5 Å) Sequential and Medium-Range ¹H–¹H Distances in Polypeptide Secondary Structures

Distance	α-helix	3_{10}-helix	β	β_P	turn I[a]	turn II[a]
$d_{\alpha N}$	3.5	3.4	2.2	2.2	3.4	2.2
					3.2	3.2
$d_{\alpha N}(i,i+2)$	4.4	3.8			3.6	3.3
$d_{\alpha N}(i,i+3)$	3.4	3.3			3.1-4.2	3.8-4.7
$d_{\alpha N}(i,i+4)$	4.2					
d_{NN}	2.8	2.6	4.3	4.2	2.6	4.5
					2.4	2.4
$d_{NN}(i,i+2)$	4.2	4.1			3.8	4.3
$d_{\beta N}$[b]	2.5-4.1	2.9-4.4	3.2-4.5	3.7-4.7	2.9-4.4	3.6-4.6
					3.6-4.6	3.6-4.6
$d_{\alpha\beta}(i,i+3)$[b]	2.5-4.4	3.1-5.1				

[a] For the turns, the first of two numbers applies to the distance between residues 2 and 3, the second to that between residues 3 and 4 (Fig. 7.12). The range indicated for $d_{\alpha N}(i,i + 3)$ corresponds to the distances adopted if ψ_1 is varied between -180 and $180°$.
[b] The ranges given correspond to the distances adopted by a β-methine proton if χ^1 is varied between -180 and $180°$.

7.5. IMPLICATIONS FOR THE ANALYSIS OF NOESY SPECTRA

In ^{1}H NOESY spectra of proteins the intraresidue cross peaks between geminal and vicinal protons can be identified by comparison with the corresponding ^{1}H COSY spectrum, since these protons are also connected by scalar spin–spin coupling. In the spectral regions of prime interest for structural studies, a further large percentage of the cross peaks can usually be identified in a thorough search for sequential connectivities. The residual spectrum typically contains a relatively small number of cross peaks manifesting short medium-range and long-range ^{1}H–^{1}H distances. In Figure 7.14

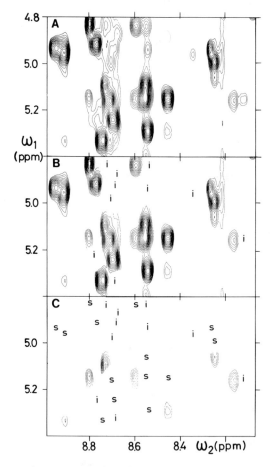

Figure 7.14. Intraresidue, sequential, and medium-range or long-range cross peaks in the NH–αH region of the ^{1}H NOESY spectrum of Tendamistat (0.007 M, H$_2$O, pH 3.2, 50 °C; 500 MHz; τ_m = 200 ms). (A) Complete spectrum. (B) Same as (A), intraresidue cross peaks replaced by i. (C) Same as (B), sequential cross peaks replaced by s. The remaining peaks correspond to long-range backbone connectivities $d_{\alpha N}(i,j)$ between neighboring strands in the antiparallel β sheets (see text).

this is illustrated for part of the NH–αH cross-peak region in the ^{1}H NOESY spectrum of Tendamistat. The secondary structure in this protein consists of antiparallel β sheets and tight turns, with complete absence of helical segments (Kline and Wüthrich, 1985). Therefore nearly all residual peaks in Figure 7.14C correspond to long-range backbone distances $d_{\alpha N}(i,j)$. In α-helical proteins, there would instead be residual cross peaks manifesting medium-range ^{1}H–^{1}H distances $d_{\alpha N}(i,i + 3)$ and $d_{\alpha N}(i,i + 4)$ (Williamson et al., 1984; Zuiderweg et al., 1983a).

CHAPTER 8
Sequence-Specific Resonance Assignments in Proteins

This chapter describes procedures for assigning the ^{1}H NMR lines of all (or in practice nearly all) hydrogen atoms in small proteins, which was already briefly alluded to in Section 4.1. In a polypeptide chain of 100 amino acid residues, assignments must be obtained for approximately 400–700 nonlabile hydrogen atoms, and 110–140 potentially NMR-observable, labile protons. The general strategy used to solve this problem includes several different steps of resonance identification and resonance assignment (Fig. 8.1):

1. The *spin systems* of the nonlabile protons in individual amino acid residues are identified in D_2O solution of the *native* protein, using as far as possible (Table 2.2) through-bond ^{1}H–^{1}H connectivities (Section 5.5).

2. From further studies in H_2O solution of the native protein, the amino acid spin systems of the individual residues are completed through identification of the *J* connectivities with the labile protons.

3. *Sequentially neighboring* amino acid ^{1}H spin systems are identified from observation of the sequential NOE connectivities $d_{\alpha N}$ or d_{NN}, or possibly $d_{\beta N}$.

4. The aim of the experimental steps *1–3* is to identify groups of ^{1}H NMR lines corresponding to *peptide segments* that are sufficiently long to be *unique in the primary structure* of the protein. Sequence-specific assignments are then obtained by matching the peptide segments thus identified by NMR with the corresponding segments in the *independently determined* (*not* by NMR) amino acid sequence.

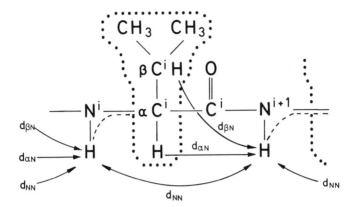

Figure 8.1. Polypeptide segment with indication of the spin systems of nonlabile protons in the individual residues (inside dotted lines), the αH–NH COSY connectivities (broken lines), and the sequential NOE connectivities (arrows).

An illustrative example for this general strategy is provided by the DNA-binding domain 1-51 of *E. coli lac* repressor (*lac* headpiece). The sequence of 51 residues includes three unique amino acids, Gly-14, His-29, and Ile-48, and numerous dipeptide segments with unique NMR properties (Fig. 8.2). For unique residues the sequence-specific assignment follows directly from the identification of the spin system. For the resonances of the unique dipeptides, sequence-specific assignments result if a sequential connectivity between the ^{1}H spin systems of the two residues can be established. The unique residues and dipeptide segments can then be used as reference locations in the primary structure, so that further sequential identification steps directly provide the sequence-specific assignments for the adjacent residues.

The through-bond connectivities used to delineate the ^{1}H spin systems of the individual residues are unambiguous in the sense that no connectivities

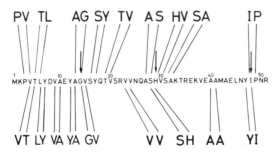

Figure 8.2. Unique amino acid residues (arrows) and selected unique dipeptide segments (large letters) in the *lac* headpiece, which are suitable references for obtaining sequence-specific ^{1}H NMR assignments.

Figure 8.3. Short nonsequential ^{1}H–^{1}H distances in proteins.

between protons located in different residues are to be expected. In contrast, the identification of neighboring residues by sequential NOE connectivities is not a priori unambiguous. As a consequence of the spatial folding of the polypeptide chain, short distances $d_{\alpha N}(i,j)$, $d_{NN}(i,j)$, and $d_{\beta N}(i,j)$ may also prevail between nonneighboring residues. (In Fig. 8.3 such long-range short contacts are indicated by dotted circles.)

An estimate of the reliability of sequential assignments by NOE's was obtained from a statistical analysis of short ^{1}H–^{1}H distances in a group of high resolution protein crystal structures (Billeter et al., 1982). Table 8.1 shows, for example, that in these crystal structures 98% of the αH–NH distances shorter than 2.4 Å are between sequentially neighboring residues, and that this percentage drops to 72% when the distance limit is increased to 3.6 Å. Similar, though somewhat lower percentages of sequential connectivities prevail for short NH–NH and βH–NH distances. Consistently high percentages of sequential connectivities result when short distance limits are imposed simultaneously for two such distances (three examples are listed in Table 8.1). For practical purposes, these statistical data indicate that in each group of 10 NH–NH, αH–NH, or βH–NH NOESY cross peaks recorded with a mixing time of approximately 100 ms, one or two connectivities are between nonneighboring residues and would lead to erroneous results if they were used for sequential assignments. These potential errors must be eliminated by checking the sequential assignments for consistency with the amino acid sequence determined by chemical or biochemical methods.

In view of the inherent 10–20% error rate in the initial identification of sequentially neighboring spin systems by NOE's, and considering that these errors could accumulate in the course of the assignment procedure, the overall reliability of the assignments depends critically on the frequency of unique short peptide segments in the amino acid sequence. To assess this important factor, a large sample of protein sequences shorter than 200 residues was screened for repeats of identical peptide segments. The probability that a predetermined dipeptide segment occurs only once in a protein was thus found to be 56%, and for tri- and tetrapeptide segments this probability is, respectively, 95 and 99% (Table 8.2). In practice, all amino acid types

TABLE 8.1. Statistics of Short ^{1}H–^{1}H Distances in Protein Crystal Structures.[a]

Distance (Å)		$j-i = 1$ (%)
$d_{\alpha N}(i,j)$	$\leqslant 2.4$	98
	$\leqslant 3.0$	88
	$\leqslant 3.6$	72
$d_{NN}(i,j)$	$\leqslant 2.4$	94
	$\leqslant 3.0$	88
	$\leqslant 3.6$	76
$d_{\beta N}(i,j)$	$\leqslant 2.4$	79
	$\leqslant 3.0$	76
	$\leqslant 3.6$	66
$d_{\alpha N}(i,j) \leqslant 3.6,\ d_{NN} \leqslant 3.0$		99
$d_{\alpha N}(i,j) \leqslant 3.6,\ d_{\beta N} \leqslant 3.4$		95
$d_{NN}(i,j) \leqslant 3.0,\ d_{\beta N} \leqslant 3.0$		90

[a] The numbers indicate which percentage of the distances identified in the first column are observed between sequentially neighboring residues. A sample of 19 proteins with 3224 amino acid residues was investigated (Billeter et al., 1982).

TABLE 8.2. Probability That a Predetermined Di-, Tri-, or Tetrapeptide Segment Is Unique in a Protein with Less Than 200 Residues[a]

Number of Amino Acid Types	Probability of Uniqueness (%)		
	Dipeptide	Tripeptide	Tetrapeptide
18[b]	56	95	99
8[c]	15	53	83
13[d]	33	78	95
15[e]	45	91	98

[a] Statistical data on 1905 amino acid sequences of globular proteins shorter than 200 residues taken from the protein-sequence data base of the National Biomedical Research Foundation, Washington, D.C., USA.
[b] Common L-amino acids, with Asx = Asp + Asn, Glx = Glu + Gln, since these are often listed as Asx or Glx in the data base.
[c] Gly, Ala, Val, Leu, Ile, Thr, □ = all AMX spin systems of αCH–βCH$_2$, ■ = all others (long side chains).
[d] Same as footnote c, except that the eight AMX species are further distinguished as follows: Phe, Tyr, Trp, His, Ser, ◨ = Cys, Asp, and Asn.
[e] Same as footnote d, except that the following three subgroups of long side chains are distinguished: Pro; ▼ = Lys and Arg; ▲ = Met, Glu, and Gln.

cannot readily be distinguished from their ¹H spin systems (Table 2.2). We have therefore further examined the situations arising from three different levels of spin system identification, where the 20 amino acids are divided up into 8, 13, or 15 different species (Table 8.2). On the lowest level considered, there are the 6 amino acids with unique side chain symmetry, a group comprising the 8 αCH–βCH₂ AMX spin systems, and a group containing 6 *long side chains*. On the higher levels of spin system identification, the amino acids in the latter two groups are further distinguished as specified in the footnotes to Table 8.2. As an illustration, Figure 8.4 shows the sequence of glucagon on the levels of 8 and 15 different amino acid species. Two important conclusions can be drawn from the results of this statistical analysis (Table 8.2). First, unique tri- or tetrapeptide segments are ubiquitous in globular proteins, indicating that the assignment procedure described here is generally applicable. Second, for a given protein the reliability of the sequential assignments is greatly improved if more specific spin system identifications can be obtained. In glucagon, for example, there are 9 unique dipeptides on the level of distinguishing between 8 amino acid species, whereas 24 out of the total of 28 dipeptides are unique on the level of 15 species (Fig. 8.4). (In Fig. 8.2 those dipeptide segments of *lac* headpiece are singled out that are unique according to the following criteria, which are representative of current practice in our laboratory for the identification of reference locations for sequence-specific assignments: Spin system identification on the level of 15 species as specified in Table 8.2. Both residues in the dipeptide must be identified as unique residue types.) These considerations further indicate that it may in certain instances be attractive to create additional unique spin systems in a polypeptide chain through isotope labeling.

So far the spectral analyses leading to resonance assignments were performed manually. It is to be expected that partial or even complete automation will be used in the future, since the actual work consists primarily of recognizing simple geometric patterns in the 2D NMR spectra (Meier et al., 1984; Neidig et al., 1984; Pfändler et al., 1985). It may be added that use of ¹H–¹H NOE's for obtaining sequential assignments is intrinsically an efficient method, since the $d_{\alpha N}$, d_{NN}, and $d_{\beta N}$ NOESY cross peaks would eventually have to be identified in the course of the spatial-structure determina-

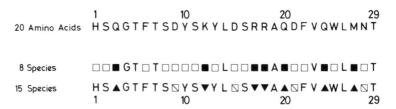

Figure 8.4. Amino acid sequence of glucagon on the levels of the 20 common amino acids, 8 different residue types, and 15 different residue types (the symbols used are explained in the footnotes to Table 8.2).

tion (Fig. 7.14), even if the sequential connectivities were obtained by a different technique.

In principle, sequence-specific resonance assignments may to some extent be obtained with 1D NMR experiments (Dubs et al., 1979; Inagaki et al., 1981). However, this is of very limited practical interest and therefore the following discussion exclusively considers the use of 2D NMR methods.

8.1. IDENTIFICATION OF THE AMINO ACID ¹H SPIN SYSTEMS

To identify the ¹H spin systems of the individual amino acid residues in a protein, one will normally start with ¹H COSY in D_2O solution after replacement of all labile protons with deuterium. Typically, for a small protein such COSY spectra contain several hundred resonance peaks. In the spectral analysis one identifies those groups of J-coupled resonances all of which arise from the same amino acid residue. This amounts to recognizing particular geometric patterns in the COSY spectra, which are characteristic for the different spin systems encountered in the 20 common amino acid residues (Table 2.2). A survey of these COSY connectivity patterns is given in Figure 8.5. For example, Gly and Ala give rise to one COSY cross peak, Val has a pattern containing two $\gamma CH_3-\beta H$ cross peaks and a $\beta H-\alpha H$ cross peak, and AMX spin systems produce another characteristic motive of three cross peaks. In all, the 20 common amino acid residues give rise to 10 different COSY connectivity patterns for the aliphatic protons (Fig. 8.5A and B), and four motives for the aromatic rings (Fig. 8.5C). To a certain extent these coupling patterns can be recognized in low resolution, absolute value COSY spectra of proteins. For example, below the diagonal in the spectrum of micelle-bound glucagon (Fig. 8.6) the connectivities for four complete AMX spin systems are identified by solid or broken lines. Above the diagonal, three different complete AMX spin systems are outlined. For two additional systems, identified by S, only one cross peak is seen, since the two β protons are degenerate. Often one finds that the two $\alpha H-\beta H$ cross peaks in $\alpha CH-\beta CH_2$ fragments have unequal intensities, which reflects that one of the $\alpha H-\beta H$ coupling constants is larger than the second one (Fig. 5.33).

Information corresponding to that from COSY can also be obtained from SECSY. Actually, for the reasons outlined in Section 5.5, SECSY was the preferred technique for spin system identification in most of the early investigations of proteins (e.g., Arseniev et al., 1982, 1983; Keller et al., 1983; Nagayama and Wüthrich, 1981a; Štrop et al., 1983a). SECSY connectivity patterns for selected residues are shown in Figure 8.7. (Complete listings for the aliphatic spin systems are presented in Nagayama and Wüthrich, 1981a, and for the aromatic spin systems in Štrop et al., 1983a.) The geometric arrangement of the connectivities for different amino acids can readily be derived from the general outlay of SECSY spectra in Figure 5.6. Similar to the analysis of COSY, information on the spin systems is obtained through a

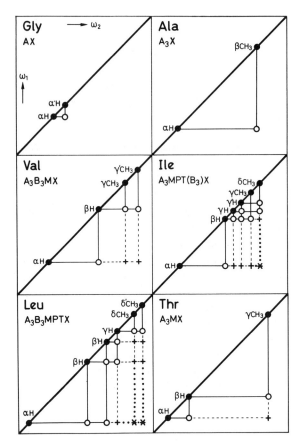

Figure 8.5. COSY, RELAYED-COSY, and DOUBLE-RELAYED-COSY connectivity diagrams for the spin systems of nonlabile protons in the common amino acid residues (Table 2.2). Each square represents a 2D spectrum for an amino acid residue or a group of residues. The positions of the diagonal peaks are represented by filled circles, COSY connectivities by open circles and solid lines, RELAYED-COSY by crosses and broken lines, DOUBLE-RELAYED-COSY by stars and dotted lines. Additional relayed connectivities may be propagated through the geminal coupling in CH_2 groups, for example, $\alpha H \rightarrow \beta H \rightarrow \beta' H$. Arbitrary chemical shifts have been chosen to enable a clear presentation.

search for these connectivity patterns in the SECSY spectrum. In Figure 8.8, solid lines indicate the complete connectivity patterns for one valyl and three threonyl residues. [For two of these Thr the relative order of the αH and βH chemical shifts is inverted relative to the random coil values (Table 2.3). This illustrates again that because of the conformation dependence of the chemical shifts, the spin system identification must rely primarily on recognition of the molecular symmetry.] For one Ser, the three chemical shift positions are indicated by arrows; the complete set of connectivities for this AMX spin system (Fig. 8.7) can be observed. For each of the seven Ala,

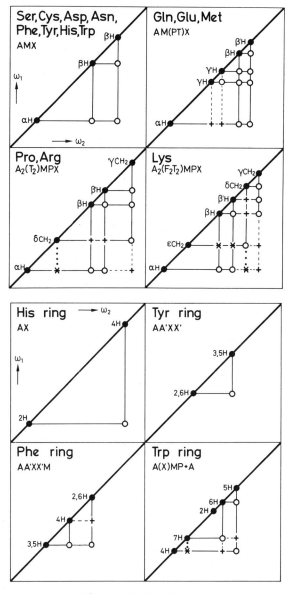

Figure 8.5. (Continued).

one βCH_3–αH cross peak is identified, which is sufficient to characterize the entire spin system (Fig. 8.7).

In the early protein ^{1}H NMR assignments, the initial spin system identifications relied almost entirely on observation of the connectivity patterns in absolute value SECSY and COSY spectra, with some additional information on spin–spin coupling fine structure coming from 2D J-resolved spectra

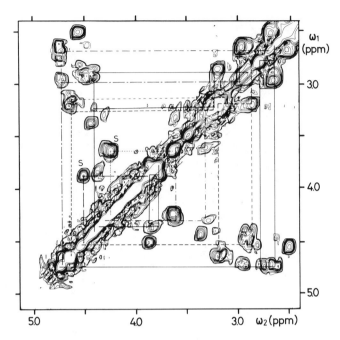

Figure 8.6. Identification of AMX and A_2X spin systems in an absolute value ^1H COSY spectrum of micelle-bound glucagon (same conditions as in Fig. 5.11, except that the solvent was D_2O; digital resolution 5.8 Hz/point) (adapted from Wider et al., 1982).

(Nagayama and Wüthrich, 1981a). With these techniques the analysis was carried approximately to the level of distinguishing between 8 different amino acid species, as specified in Table 8.2. Further discrimination was then obtained only in conjunction with the sequential assignments, as will be further discussed in Sections 8.2 and 8.3. Presently, the spin system identifications prior to sequential assignments can be pursued to a level of distinguishing between 12 to 18 species among the 20 amino acid residues. This relies on using, in addition to COSY, RELAYED-COSY and DOUBLE-RELAYED-COSY (Fig. 8.5), MQ spectroscopy, TOCSY, and NOESY recorded in D_2O and H_2O (Fig. 7.5). In the following we consider briefly the information obtained with these techniques for the individual amino acid residues.

In proteins the αCH_2 group of Gly usually gives an AX spectrum with an outstandingly large doublet splitting arising from the geminal coupling constant of 15 Hz. From the high intensity and the partially resolved fine structure, the glycines can usually be unambiguously recognized even with absolute value COSY or SECSY (e.g., Nagayama and Wüthrich, 1981a). In phase-sensitive COSY they are readily identified as intense, antiphase square arrays (Fig. 5.28) with large line separation (Fig. 8.9). However, Gly

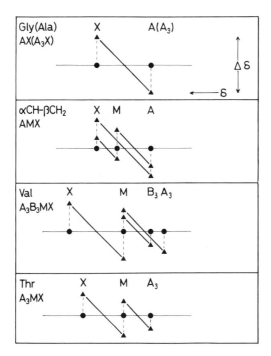

Figure 8.7. SECSY connectivity diagrams for some amino acid spin systems. The peaks corresponding to the diagonal peaks in COSY are represented by filled circles, cross peaks by filled triangles. The distance of a cross peak between spins A and X from the central, horizontal axis is $\frac{1}{2}|\Delta\delta| = \frac{1}{2}|\delta_X - \delta_A|$ (from Nagayama and Wüthrich, 1981a).

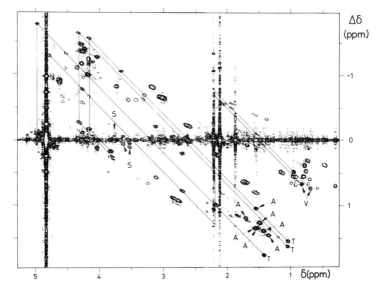

Figure 8.8. Identification of amino acid spin systems in a ¹H SECSY spectrum of *lac* headpiece (0.005 M, D_2O, pD 5.1, 18°C; 500 MHz; absolute value; digital resolution 2.2 Hz/point) (from Zuiderweg et al., 1983b).

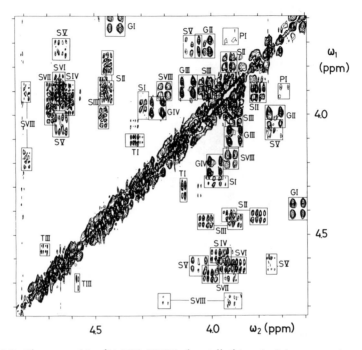

Figure 8.9. Phase-sensitive ^{1}H 2QF-COSY of metallothionein-2 (same spectrum as Fig. 5.18). The spectral region shown contains the connectivities αH–αH for Gly, αH–βH for Ser and Thr, βH–βH for Ser, and δH–δH for Pro. Positive and negative levels are plotted without distinction. The cross peaks in boxes are identified by residue type and an arbitrary Roman numeral index. Some additional, weaker peaks are due to impurities (adapted from Neuhaus et al., 1985).

may escape detection in DQF-COSY in D_2O if the two α protons have identical or nearly identical chemical shifts. An unambiguous identification must then rely on spectra recorded in H_2O, and can be based, for example, on the fine structure of the NH–αCH$_2$ COSY cross peak, or on observation of an equivalence peak at the position $[\omega_2 = \delta(NH), \omega_{1rel} = 2\delta(\alpha H)]$ in a 2Q spectrum (Fig. 5.25) (Otting and Wüthrich, 1986; Wagner and Zuiderweg, 1983).

The identification of the αCH–βCH$_2$ AMX-spin systems by COSY is in principle straightforward (Figs. 5.28 and 8.5*B*). For example, in Figure 8.9, the four seryl residues SII, SIII, SV, and SVIII all show the complete AMX connectivities (Fig. 5.28). Ambiguities may arise when the two β protons have nearly identical chemical shifts, since the two αH–βH cross peaks are then overlapped and the βH–β'H cross peak is covered by the diagonal (seryl residues SI, SIV, SVI, and SVII in Fig. 8.9). Incomplete AMX patterns can also result when one of the αH–βH spin–spin couplings is very small and gives rise to a peak with small intensity (Fig. 5.33). Furthermore, typical proteins contain a large number of AMX residues, so that the spectral regions containing these cross peaks are very crowded. Careful analysis of the fine structure in the resolved cross peaks can often lead to a reconstruction of the missing spectral features. Remaining difficulties can in most instances be resolved by supplementing the COSY experiments with RELAYED-COSY. A typical relay pathway is from αH to the β proton with the larger αH–βH coupling, and from there via the geminal coupling to the second βH. In spectra with overlap in the αH region, DOUBLE-RELAYED-COSY in H_2O can make use of the generally larger separation of the NH chemical shifts in the identification of both βH resonances (an example is shown in Fig. 5.23). In situations of degeneracy or near degeneracy of two resonances, 2Q spectroscopy may be helpful, since there is no interference with a diagonal (Figs. 5.25 and 5.26). Finally, a stringent test for distinguishing AMX spin systems from those of long side chains is that the βH–βH cross peaks of AMX should be completely suppressed in a 4QF-COSY spectrum or a 4Q spectrum in D_2O.

Once the AMX spin systems are identified, further differentiation into the eight amino acids that contain an αCH–βCH$_2$ fragment is important. Practical experience indicates that with the possible exception of some histidines, the four aromatic side chains can reliably be identified from observation of the short βH-ring proton distances (Fig. 7.5) by NOESY, provided that both the AMX systems and the aromatic spin systems were independently identified. As an illustration, Figure 8.10 shows the spectra used for identification of the aromatic residues in Tendamistat. Horizontal lines connect NOESY cross peaks with aromatic protons on the left to COSY cross peaks of AMX spin systems at the positions $[\omega_1 = \delta(\beta)$ or $\delta(\beta'), \omega_2 = \delta(\alpha)]$ on the right. These experiments further present a criterion for the identification of the 2,6-proton resonances in Phe and Tyr and the 4H line in Trp. The Ser spin systems can be identified from the outstandingly small geminal coupling, $^2J_{\beta\beta'} \approx 12$ Hz, which is in most cases corroborated by the low field positions of the βH lines. Asn can in principle be identified from the strong βCH–δNH cross peaks in NOESY spectra recorded in H_2O (Section 7.2), and Asp from the pH dependence of the βH chemical shifts (Nagayama and Wüthrich, 1981a). In practice, however, the distinction between Asp, Asn, and Cys has so far in most cases been postponed to a later stage of the assignment procedure.

Ala and Thr both give strong cross peaks in similar locations, which connect methyl doublets to protons at lower field (Figs. 5.28 and 5.31). A distinction between Ala and Thr can rest on proving the presence of coupling to a third type of proton in the case of the threonine spin systems. The differing effects of overlap on in-phase and antiphase splittings (Section 5.6) imply that even if $^3J_{\alpha\beta}$ for a threonine is too small to be resolved as an in-phase splitting of the γCH$_3$–βH cross peak, the αH–βH cross peak, in which $^3J_{\alpha\beta}$ appears in antiphase, may still be detectable (an example is TIII in Figs. 5.31 and 8.9). Independently, RELAYED-COSY experiments in H_2O show

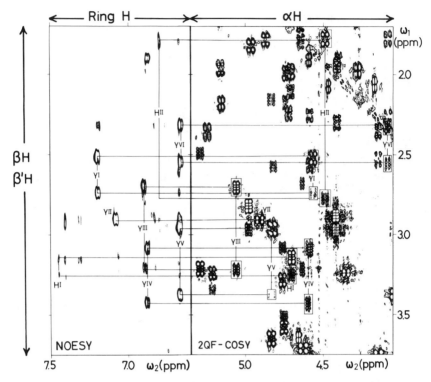

Figure 8.10. Regions from phase-sensitive 2QF-COSY and NOESY spectra of Tendamistat used for the resonance identifications in aromatic side chains (0.005 M, D_2O, pD 3.2, 50°C; 500 MHz; τ_m = 200 ms). Horizontal lines connect the $\alpha H–\beta H$ and $\alpha H–\beta'H$ cross peaks in COSY with the intraresidue cross peaks between ring protons and βH and $\beta'H$ in NOESY. The individual connectivities are identified with the amino acid type and an arbitrary Roman numeral index (adapted from Kline and Wüthrich, 1986).

cross peaks only for the methyls of alanine. In principle the two spin systems can also be distinguished with experiments using 4QF-COSY and 5QF-COSY, or 4Q and 5Q spectra. [As yet, because of low sensitivity, these latter experiments are hardly practicable with proteins, except possibly 5QF-1D NMR. See Rance et al. (1985c).] Overall, while identification of Ala and Thr appears in principle straightforward, difficulties usually arise in distinguishing Ala spin systems from Thr systems with degenerate αH and βH chemical shifts, or with a very small coupling constant $^3J_{\alpha\beta}$.

For Val, Leu, and Ile the methyl resonances and the αH resonances are usually well separated, except that in proteins containing more than approximately 12 of these residues, the methyl region may be rather crowded. Even though in several proteins complete spin system identifications for the valines were obtained with absolute-value SECSY, the experiments to be used are RELAYED-COSY for Val and Ile, DOUBLE-RELAYED-COSY for Leu and Ile (Fig. 8.5A), or TOCSY. Direct connectivities from αH to the methyl groups can thus be established even in situations, where all the intervening proton lines cannot be resolved because of the high multiplicities, strong-coupling effects, or mutual overlap. Direct evidence for distinguishing δCH₃ peaks of Ile from the other methyl peaks of Val, Leu, and Ile can be obtained from the COSY cross-peak fine structure.

The five-spin systems of Glu, Gln, and Met can usually be distinguished from the other long side chains by a combination of arguments, including complete delineation of the spin–spin coupling network, analysis of the fine structure in the cross peaks with the γ protons, and the fact that the random coil chemical shifts for γCH₂ are at lower field than for Pro, Lys, and Arg (Table 2.3). In principle, Glu can be individually identified from the pH-dependent chemical shift of the γ protons (Nagayama and Wüthrich, 1981a) and Gln from the strong γCH–εNH cross peaks in NOESY spectra recorded in H₂O (Section 7.2). Usually these distinctions have been postponed to later stages of the assignment procedure. In proteins containing two or several methionines, there is usually no direct, spectroscopic technique available for combining the εCH₃ resonances with the corresponding five-proton systems.

Pro can be distinguished from the other long side chains by the absence of a NH–αH cross peak in COSY spectra recorded in H₂O. In many instances, complete assignments for Pro were obtained, often in conjunction with the sequential assignments (Kline and Wüthrich, 1986; Neuhaus et al., 1985; Štrop et al., 1983a; Wagner and Wüthrich, 1982a; Wagner et al., 1986).

With few exceptions (Brown and Wüthrich, 1981; Kline and Wüthrich, 1986), the spin systems of Lys and Arg have not been identified beyond βCH₂. Since these long, polar side chains are predominantly located on the protein surface and stick out into the solvent, the dispersion of the chemical shifts is usually too small to allow separate observation of the resonances of peripheral protons in different side chains. However, recent TOCSY experiments provided triple-relayed αH-εCH₂ connectivities in Lys (G. Otting and K. Wüthrich, to be published), where the dispersion of the αH shifts enables separate observation of the entire spin systems.

For all amino acid residues the completion of the spin system by delineation of the αH–NH connectivity in H₂O solution of the protein is a very important part of the resonance identifications. Because of the limited resolution along the αH-chemical shift axis in absolute-value COSY, only incomplete sets of NH–αH connectivities could be obtained in the initial assignments of protein ¹H NMR spectra (e.g., Arseniev et al., 1982; Keller et al., 1983; Wagner and Wüthrich, 1982a) and a considerable fraction of the NH–αH connectivities were then established only in conjunction with the sequential assignments (Wüthrich, 1983). With the improved resolution in phase-sensitive COSY fingerprints and using RELAYED-COSY and 2Q spectroscopy, essentially complete identification of the NH–αH COSY cross peaks can now be obtained prior to the sequential assignments (Kline and Wüthrich, 1986; Neuhaus et al., 1985). Figure 8.11 illustrates the power

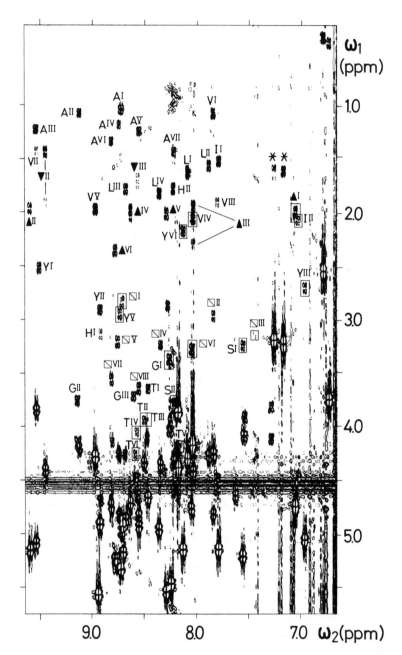

Figure 8.11. Phase-sensitive RELAYED-COSY spectrum of Tendamistat (0.005 *M*, H$_2$O, pH 3.2, 50°C; 500 MHz, $\tau = 30$ ms) used for connecting the amino acid side chain spin systems with the backbone amide protons. Relayed connectivities are identified with the residue type (spin system identifications had been carried to the level of 15 species as specified in Table 8.2; the stars identify peaks originating from side chain protons) and an arbitrary Roman numeral index (adapted from Kline and Wüthrich, 1986).

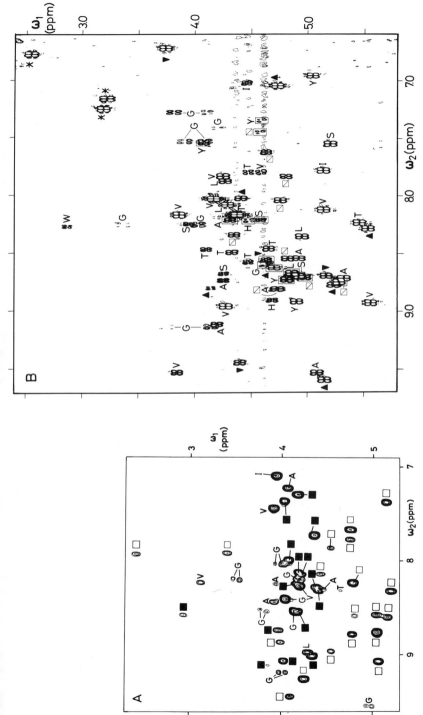

Figure 8.12. COSY fingerprints of two proteins with spin system identifications. (A) Absolute value COSY of BUSI (0.012 M, H₂O, pH 5.5, 45°C; 500 MHz). The peaks are labeled with one-letter symbols for the amino acid types, corresponding to spin system identifications on the level of 8 species as specified in Table 8.2. (B) Phase-sensitive DQF-COSY of Tendamistat (same solution as Fig. 8.11; 500 MHz) with peak identifications after discriminating between 15 residue types. The stars identify side chain cross peaks from Arg.

of RELAYED-COSY for this step of the spectral analysis. It is particularly worth noting that the resolution of the relayed connectivity cross peaks along ω_1 between 1.0 and 4.3 ppm is largely improved when compared with the corresponding direct cross peaks in the COSY-fingerprint region between $\omega_1 = 3.8$ and 5.8 ppm.

As a survey of the results obtained after the steps *1* and *2* of the assignment procedure, Figure 8.12 shows the COSY fingerprints of two proteins with spin system identifications. In the absolute-value spectrum of BUSI (Fig. 8.12*A*), only 8 different groups of amino acids were distinguished, as specified in Table 8.2, whereas in Tendamistat (Fig. 8.12*B*) 15 different species were distinguished. The symbols used in Figure 8.12 for the identification of amino acid types are defined in the footnotes to Table 8.2.

8.2. SEQUENTIAL ASSIGNMENTS VIA ^{1}H–^{1}H OVERHAUSER EFFECTS

The information needed for obtaining the sequential connectivities between neighboring spin systems is obtained from analysis of the regions containing the NH–αH, NH–NH, and NH–βH cross peaks in NOESY spectra recorded in H_2O (Fig. 6.8, regions a, b, and c). In practice, however, one needs both NOESY and COSY, and suitable combination and visual display of information on through-space and through-bond connectivities is indispensable.

The spectral region of interest for the sequential assignments corresponds approximately to the area ($\omega_1 = 1.0$–11.0 ppm, $\omega_2 = 7.0$–11.0 ppm). All three sequential connectivities involve at least one potentially exchangeable amide proton (Fig. 8.1). A complete set of sequential assignments can therefore only be obtained from NOESY spectra recorded in H_2O under conditions of pH and temperature where the exchange with the solvent is sufficiently slow to allow observation of all backbone amide protons (Sections 2.3, 3.4, and 5.7). In many proteins it may, however, be helpful to work also with subspectra containing only part of the NH lines. Typically this would include a subspectrum recorded in a freshly prepared D_2O solution, which contains only the slowly exchanging amide-protons (Fig. 8.13), and a subspectrum recorded in a H_2O solution freshly prepared after the slowly exchanging NH's were replaced by deuterium, which contains only the rapidly exchanging amide protons (Keller et al., 1983).

Inspection of Figure 8.1 shows that a continuous pathway along the backbone of a peptide segment is obtained using alternatively COSY connectivities NH$_i$–αH$_i$ and NOESY connectivities $d_{\alpha N}$. A clear presentation is afforded by NOESY–COSY connectivity diagrams (Wagner et al., 1981). These make use of the inherent symmetry of COSY and NOESY spectra with respect to the diagonal. The upper left triangle of a NOESY spectrum is combined with the lower right triangle of a COSY spectrum recorded from the same solution (Fig. 8.13). (In practice it may be useful to transpose the

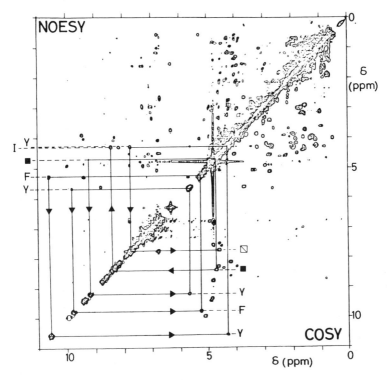

Figure 8.13. NOESY–COSY connectivity diagram for sequential ¹H NMR assignments via $d_{\alpha N}$ in a freshly prepared D_2O solution of BPTI (0.02 M, pD 4.6, 24°C; 360 MHz; absolute value; τ_m = 100 ms). The assignment pathway for a segment of six residues is indicated. The symbols for amino acid types are explained in Table 8.2 (adapted from Wagner et al., 1981).

upper left triangle of the COSY spectrum into the lower right triangle by a modification of the plot routine. By this one prevents loss of information from overlap with the spurious signals at the ω_2 position of the solvent resonance.) Analysis of the NOESY–COSY connectivity diagram starts with a COSY cross peak, for example, in Figure 8.13 with the cross peak attributed to a long side chain (in this study the spin system identification was carried to the level of distinguishing between 13 species, as specified in Table 8.2). A vertical and a horizontal line through the COSY cross peak disect the diagonal at the αH–chemical shift and the NH-chemical shift of this residue, respectively. We start assignments in the direction toward the C terminus (Fig. 8.1). The $d_{\alpha N}$ cross peak connecting αH of the starting residue with NH of the following residue must lie in the NOESY spectrum on the horizontal line through the diagonal peak of this αH. In Figure 8.13 there is only one NOESY cross peak that satisfies this condition, which is therefore taken to be the desired connection. A vertical line leads from this cross peak to the diagonal NH position, which contains in this case a well separated resonance peak. Continuing on, a horizontal line into the COSY spectrum leads to the cross peak of a Tyr. We have now performed one complete assignment cycle in the NOESY–COSY connectivity diagram, with the result that the resonance lines of a dipeptide segment ■Y are identified. (Here and in the following, the symbols ■, □, ◨, ▲, and ▼ represent a residue from a group of residues with similar NMR properties, as specified in Table 8.2.) The second complete assignment cycle results in specifying the resonances of a tripeptide ■YF, and after a total of four cycles in the connectivity diagram we find the pentapeptide sequence ■YFY◨. Starting again from the COSY cross peak of ■, we can also pursue sequential assignments in the opposite direction, toward the N terminus. The $d_{\alpha N}$ NOESY cross peak to the preceding residue must lie on a vertical line through the diagonal NH position of residue ■. The single peak that satisfies this condition identifies the αH-chemical shift of the preceding residue at 4.35 ppm. The subsequent search for the αH_i–NH_i COSY cross peak yields no result, so that the assignment pattern ends (this connectivity was later on found in spectra recorded in H_2O). Since in BPTI there are two spin systems with an identical αH-chemical shift of 4.35 ppm, namely, Ile and Ala, the residue type could not be uniquely determined at this stage, so that overall the result of the analysis indicated in Figure 8.13 is that the ^{1}H NMR lines of a peptide segment (A/I)■YFY◨ were identified.

Figure 8.14 shows a different presentation of a NOESY–COSY connectivity diagram. Identical spectral regions containing the NH–αH cross peaks in NOESY and COSY are plotted on an expanded scale and joined at the nearest corners. The main advantage of this is that more details can be recognized in the relevant spectral regions. To establish sequential connectivities, the same rules apply as for Figure 8.13. The sequential assignments for two peptide segments are outlined. The first segment starts from the COSY cross peak of a Leu, which is identified by a dot in its center. A vertical line through this peak disects the pseudodiagonal at the Leu αH-chemical shift, and the $d_{\alpha N}$ connectivity to the following residue must lie on a horizontal line through this αH-chemical shift position in the NOESY spectrum. Since this line would go through more than one NOESY cross peak (Fig. 8.14), the assignment at this stage is not unambiguous. In practice one would pursue each of the possible assignment pathways and subsequently select the correct one from independent, additional evidence (e.g., observation of the sequential connectivities d_{NN} or $d_{\beta N}$, or checks of the peptide segments found via the NMR assignments against the amino acid sequence; see also Section 8.3). Similar to the procedures in Figure 8.13, the first assignment cycle is completed by a vertical line from the NOESY cross peak to the pseudodiagonal, and a horizontal line from there into the COSY spectrum. Continuing with another counter-clockwise cycle, the resonances of a tripeptide segment L▼I are identified. (In this case 15 species were distinguished in the spin system identifications; see Table 8.2.) The second

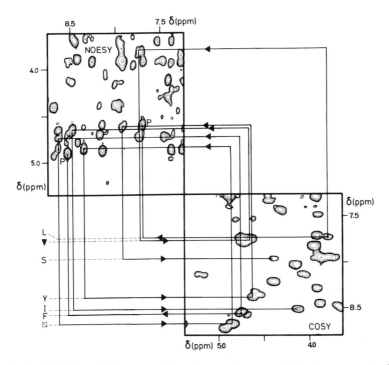

Figure 8.14. NOESY–COSY connectivity diagram for sequential assignments via $d_{\alpha N}$ in inhibitor K (0.01 M, H_2O, pH 3.4, 50°C; 500 MHz; absolute value; τ_m – 200 ms). The assignment pathways for a pentapeptide segment and a tetrapeptide segment are indicated, using the symbols for amino acid types defined in Table 8.2 (adapted from Keller et al., 1983).

assignment pathway starts with COSY of Phe and yields resonance identifications for a tetrapeptide segment F◻YS. From both, Leu and Phe, the sequential connectivity toward the C terminus leads to αH of Pro, and therefore there is no connecting COSY cross peak. Overall, the two assignment patterns indicated in Figure 8.14 thus yield resonance identifications for two peptide segments PL▼I and PF◻YS.

In an alternative presentation of sequential assignments by $d_{\alpha N}$ (Fig. 8.15), the positions of the NH–αH COSY cross peaks are marked in the NOESY spectrum. In Figure 8.15 an assignment pathway starts at the COSY cross peak position of a Ser, from which a horizontal line leads to the NOESY cross peak manifesting the connectivity to the following residue in the sequence. The assignment protocol includes the indication that $d_{\beta N}$ was also observed between the same two residues, which provided the clue for selecting this $d_{\alpha N}$ connectivity from a group of two NOESY cross peaks at the ω_1 position of the Ser COSY peak. Next, a vertical line through this NH–αH NOESY cross peak leads to the COSY cross peak of an Asp (in this

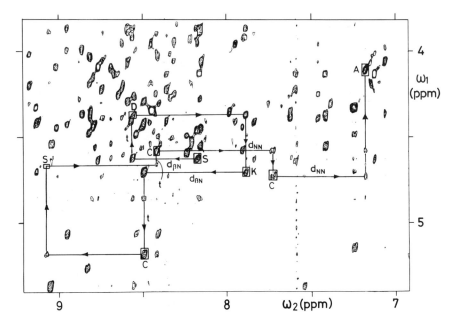

Figure 8.15. Alternative presentation of sequential assignments via $d_{\alpha N}$ in a NOESY spectrum of metallothionein-2 (0.01 M, H_2O, pH 7.0, 20°C; 500 MHz; $\tau_m = 100$ ms; absorption mode). The assignment pathway for an octapeptide segment is indicated. The squares identify the locations of the NH–αH COSY cross peaks for the residue types indicated, which coincide with the NOESY cross peaks corresponding to $d_{\alpha N}(i,i)$ (adapted from Wagner et al., 1986b).

protein, all spin systems were uniquely identified). The letter t indicates that this connectivity was unambiguous only after examination of different spectra recorded at two different temperatures, 14 and 20°C. Continuing in this fashion in the direction toward the C terminus, horizontal lines always lead from the NH–αH COSY cross-peak positions to the sequential $d_{\alpha N}$ connectivities, and vertical lines lead from the $d_{\alpha N}$ NOESY cross peaks to the NH–αH COSY cross peaks. A complete, counter-clockwise assignment cycle in a NOESY–COSY connectivity diagram (Figs. 8.13 and 8.14) corresponds in this alternative presentation to a half-cycle of one horizontal and one vertical line. Thereby the half-cycle may be clockwise or counterclockwise, depending on the relative chemical shifts of the connected protons. In Figure 8.15 the assignment pathway for an octapeptide segment SDKCSCCA is indicated. This alternative presentation of $d_{\alpha N}$ sequential assignments has been much used in the practice of spectral analysis in my group. It allows a compact survey of the observations made in different regions of the NOESY spectrum and in the COSY fingerprint, or in different spectra recorded with different conditions of temperature, pH, or ionic strength. In publications the reader may have a better chance to check on particular connectivities, since the NH–αH region of the NOESY spectrum can be presented on a more expanded scale. On the other hand, a NOESY–COSY connectivity diagram provides much simpler geometric patterns for representation of successive cycles of assignment.

For sequential assignments using d_{NN} (Fig. 8.1), a continuous series of connectivities can be displayed in the low-field region of the NOESY spectrum (Fig. 6.8, region a). In the example of Figure 8.16, the assignments started from a NOESY cross peak between Ala and Ser, which is marked with a dot. The following connectivity to the neighbor of Ala must be either on a vertical or horizontal line through the NH diagonal position, depending on whether the following NH is at higher or lower field, respectively. In Figure 8.16 it is at lower field and the residue contains a five-proton spin system, ▲ (15 species were distinguished in the spin system identifications). The following connectivity is to a residue of type ◻, which is at higher field, and there are two further cross peaks leading to another ◻ and then to ▲. The next connectivity is masked by the diagonal in the spectrum of Figure 8.16, but could be observed in a freshly prepared D_2O solution of BPTI. In

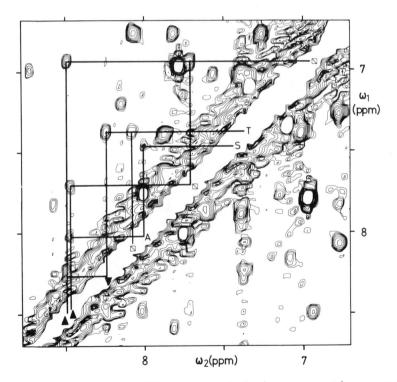

Figure 8.16. Pathway of sequential assignments via d_{NN} for a nonapeptide segment in a NOESY spectrum of BPTI (0.02 M, H_2O, pH 4.6, 68°C; 500 MHz; τ_m = 100 ms; absolute value). The symbols for amino acid types are defined in Table 8.2 (adapted from Wagner and Wüthrich, 1982a).

all, an assignment pathway for a nonapeptide is indicated in Figure 8.16. Since the chain direction is not defined by d_{NN} (Fig. 8.1), the peptide segment thus identified is either SA▲◫◫▲▼T◫ or ◫T▼▲◫◫▲AS. Additional information for distinguishing between these two possibilities can be obtained from observation of sequential connectivities $d_{\alpha N}$ or $d_{\beta N}$, or from checks against the primary protein structure.

For the presentation of sequential connectivities via $d_{\beta N}$ in Figure 8.17, information on the through-bond connectivities NH_i–αH_i–βH_i obtained from COSY was added to a NOESY spectrum in the way that the predicted locations of the intraresidue NOE's NH_i–βH_i were identified with dots. Horizontal lines lead from these intraresidue NOE's to the $d_{\beta N}$ connectivity with the following residue, and vertical lines go from the sequential $d_{\beta N}$ cross

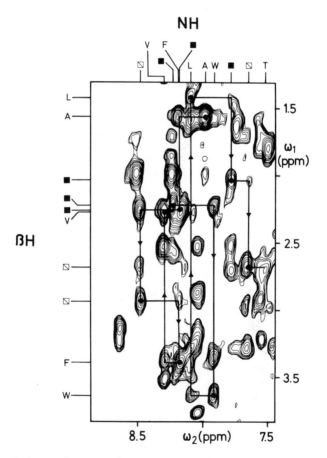

Figure 8.17. Pathway of sequential assignments via $d_{\beta N}$ for a undecapeptide segment in the NOESY spectrum of micelle-bound glucagon (same sample as Fig. 5.11; 360 MHz; $\tau_m = 100$ ms; absolute value). The symbols for amino acid types are defined in Table 8.2 (adapted from Wider et al., 1982).

peak to the intraresidue peak $d_{N\beta}(i,i)$ of this newly found residue. The assignment pathway in Figure 8.17 starts with $d_{\beta N}(i,i)$ for an alanine (13 species were distinguished in the spin system identifications, see Table 8.2) and identifies the undecapeptide segment A■◻FV■WL■◻T.

Independent of conformation, all sequential connectivities $d_{\alpha N}$ in a polypeptide chain should be observable (Fig. 7.6), so that in principle the sequential assignments might all be done with this parameter. However, Table 8.1 shows that the reliability of identification of a sequential connectivity is considerably lower when it is derived from a weak $d_{\alpha N}$ cross peak than when it comes from a strong $d_{\alpha N}$ peak manifesting an outstandingly short αH–NH distance. Therefore, two somewhat different assignment strategies may be followed. The first strategy accepts sequential connectivity information only from strong $d_{\alpha N}$ or d_{NN} peaks (with supplementary information from $d_{\beta N}$) and the assignment procedure then relies on the high intrinsic reliability of each sequential step thus obtained (Table 8.1). As an alternative, one can primarily try to obtain a complete set of $d_{\alpha N}$ connectivities. Sequential $d_{\alpha N}$ peaks must then be distinguished from αH–NH cross peaks manifesting medium-range or long-range distances of the order 3.5 Å (Table 7.1), which will again have to be based on examination of d_{NN} and $d_{\beta N}$ (Fig. 8.15).

In the practice of protein ¹H NMR assignments, difficulties tend to arise when two or several residues have identical chemical shifts for NH or αH. Since it is quite rare that cross peaks overlap in both frequency dimensions, the ensuing problems can often be solved by reference to the amino acid sequence. Thereby one makes use of the fact that once the backbone resonances are assigned to amino-acid types (Fig. 8.12), the search for sequential neighbors in each assignment step can be limited to a subgroup of cross peaks. For example, assume that the assignments start with the COSY cross peak of an Ala in a protein where only G, I, and ◻ occur as sequential neighbors of Ala, and that the sequential assignment via $d_{\alpha N}$ is ambiguous because there are two or several NOESY cross peaks with identical αH chemical shifts (Figs. 8.14 and 8.15). On the basis of the NH chemical shifts one may then single out the connectivities leading to G, I, or ◻, and can thus normally reduce the extent of the degeneracy or eliminate the ambiguity altogether. To this it may be added that the number of unassigned cross peaks and therefore the probability of encountering difficulties with chemical shift degeneracy decrease in the course of the assignment procedure.

Special consideration must be given to Pro residues. As was discussed in Chapter 7, similar conditions for obtaining sequential assignments prevail when δCH₂ of Pro is used instead of NH. The connectivities $d_{\alpha\delta}$, $d_{N\delta}$, and $d_{\beta\delta}$ are, of course, observed in different regions of the NOESY spectrum than the corresponding cross peaks with amide protons (Fig. 6.8), and $d_{\alpha\delta}$ or $d_{\beta\delta}$ are in practice best observed in NOESY spectra recorded in D₂O. The most difficult problem usually is to obtain a continuous assignment pathway across the Pro residues, since this must rely on complete assignment of the Pro spin system (Wagner and Wüthrich, 1982a). Because of the occurrence of *cis* X-Pro peptide bonds in proteins, the search for sequential assignments with Pro should also include $d_{\alpha\alpha}$ and $d_{N\alpha}$ connectivities (Fig. 7.10*B*).

8.3. SEQUENCE-SPECIFIC ASSIGNMENTS AND CHECKS ON THE PRIMARY STRUCTURE

At the outset of this chapter, Figure 8.2 was used to illustrate the principle of obtaining sequence-specific resonance assignments: The chemically determined amino acid sequence is searched for peptide segments that would match the partial sequences identified with the sequential NMR assignments. While only unique residues and dipeptides were considered in Figure 8.2, this procedure can be generalized for longer peptide segments, such as the partial sequences identified by the assignment pathways in Figures 8.13–8.17. The chemically determined amino acid sequence is needed at this stage to eliminate erroneous sequential assignment pathways, which may have resulted because of chemical shift degeneracies or because nonsequential NOE connectivities were initially interpreted as sequential ones.

The results of the sequence-specific ^{1}H NMR assignments can conveniently be presented as shown in Figure 8.18 for BUSI. In this protein the $d_{\alpha N}$ connectivities were identified between all residues except between positions 1 and 2. In addition, either one or both of the connectivities d_{NN} and $d_{\beta N}$ were observed for nearly all sequentially neighboring residues, so that high reliability for the resonance assignments was achieved on the basis of the sequential connectivities alone (Table 8.1). The data obtained from the assignment procedures also include the ^{1}H chemical shifts (Table 8.3).

In principle all the information that is missing in incomplete spin system identifications can be obtained in connection with the sequence-specific assignments. For example, different amino acid types in the groups □, ◪, ■,

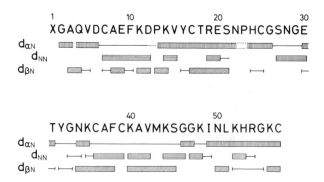

Figure 8.18. Amino acid sequence of BUSI with survey of sequential assignments. Sequential connectivities $d_{\alpha N}$, d_{NN}, and $d_{\beta N}$ manifested by strong NOESY cross peaks are indicated by shaded bands, those observed as weak NOE's by a line. The dotted lines indicate $d_{\alpha\delta}$ connectivities with Pro (from Štrop et al., 1983a).

TABLE 8.3. ^{1}H Chemical Shifts (parts per million) of the First 10 Residues in Tendamistat.[a]

Amino Acid Residue	Chemical Shifts (50°, pH 3.2)			
	NH	αH	βH	Others
Asp 1		4.42	2.98,2.90	
Thr 2	8.58	4.48	4.26	γCH$_3$ 1.23
Thr 3	8.17	4.38	4.20	γCH$_3$ 1.19
Val 4	8.03	4.17	2.06	γCH$_3$ 0.91,0.89
Ser 5	8.21	4.59	3.84,3.77	
Glu 6	8.63	4.71	2.11,1.98	γCH$_2$ 2.52,2.52
Pro 7		4.72	2.24,1.99	γCH$_2$ 1.97,1.70
				δCH$_2$ 3.76,3.65
Ala 8	8.74	4.26	1.18	
Pro 9		4.59	2.57,2.25	γCH$_2$ 2.14,2.14
				δCH$_2$ 3.68,3.41
Ser 10	8.68	4.26	4.03,3.99	

[a] From Kline and Wüthrich (1986).

▼, or ▲ (Table 8.2) can be uniquely identified once their sequence positions are determined (knowledge of the amino acid type may then often be helpful for completing the spin system identification at this stage). Or, in cases where some backbone amide-proton resonances could not be connected to the corresponding side chain spin systems by COSY or RELAYED-COSY in step **2**, this completion of the spin system identifications may result after sequential assignments from combination of all available data (Wüthrich, 1983).

Completing the spin system identifications from reference to the amino acid sequence uses the assumption that the sequence is correct. Alternatively the NMR data may eventually also allow a check on the chemical sequence determination. This is best explained using an example encountered in practice. The published sequence of BUSI that was available before the NMR assignments, contained six errors: Glu-20 and His-24 were interchanged, Asn-22 and Asn-34 were given as Asp, and Lys-56 and Cys-57 were given as Cys-56–Lys-57. The NMR correction for positions 56 and 57 was unambigous, since the two spin systems had been distinguished by COSY, and the sequential connectivities would have been incompatible with the chemically determined sequence. For residues 22 and 34, very strong NOESY cross peaks between βCH$_2$ and labile protons with amide-proton chemical shifts were observed. As long as the spatial protein structure was not known, these could in principle arise from long range-short contacts with sequentially distant residues. Therefore, before the 3D structure determination these NOESY data could be accepted as a strong implication but not as a proof for the presence of Asn side chains. A similar situation arose for the residues 20 and 24. With the NMR techniques available at the time (Štrop et al., 1983a) only the αCH–βCH₂–fragments of these two residues could be identified by through-bond connectivities. A NOESY spectrum recorded in D₂O contained cross peaks between βCH₂ of residue 24 and an imidazole ring. Before the 3D structure was known it could not be excluded that this NOE was between βCH₂ of residue 24 and the imidazole ring of His-20, so that the NMR data could again be taken as a strong indication but not as a proof of the interchange of the residues 20 and 24. The corrected sequence shown in Figure 8.18 was then firmly established by repetition of the chemical sequence determination. In another example, numerous sequence modifications in rabbit liver metallothionein-2 resulted in the course of the NMR assignments, where again the work was done in close collaboration with a laboratory specialized in sequence determination by chemical methods (Wagner et al., 1986). Two general comments may be added to this. First, exchange of sequence location between two residues attributed to the same group after spin system identification, namely, either □, ◹, ■, ▼, or ▲ (Table 8.2), would usually escape detection by NMR. Second, implications

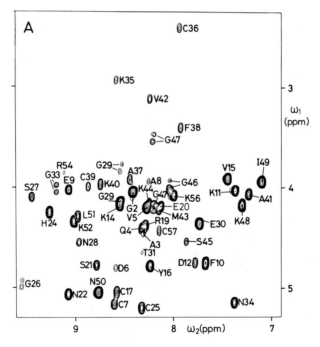

Figure 8.19. Same COSY fingerprints as in Figure 8.12 with sequence-specific resonance assignments. *(A)* BUSI (from Štrop et al., 1983). *(B)* Tendamistat. The stars identify side chain cross peaks from Arg (from Kline and Wüthrich, 1986).

for sequence errors based on NOE's can usually be accepted as firm evidence as soon as a low resolution spatial structure of the protein in solution is available.

After completion of sequence-specific resonance assignments, the peak identifications in the COSY fingerprints (Fig. 8.12) can be completed by indication of both the amino acid type and the sequence position (Fig. 8.19). Provided the NMR sample contained a homogeneous protein preparation, all NH–αH cross peaks should then be assigned, which provides an overall check on the assignment procedure.

The resonance assignments thus obtained do not include stereospecific assignments for methylene protons, the methyls in the isopropyl fragments of Val and Leu, or the aromatic protons of Phe and Tyr (Wüthrich et al., 1983). Stereospecific assignments in small peptides were obtained with heteronuclear double resonance experiments (Feeney et al., 1974) or with stereospecific isotope labeling (Fischman et al., 1978). While these techniques

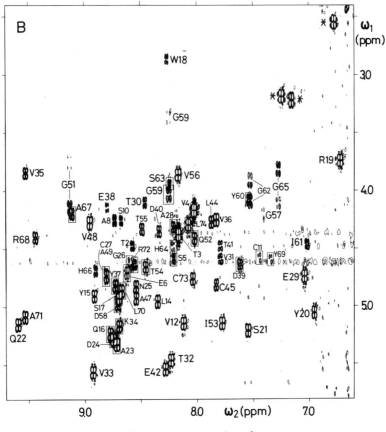

Figure 8.19. (*Continued*)

could in principle be applied to larger molecules, this appears hardly practical at the present time. In special cases, where information on the local environment of an amino acid side chain can be obtained from other experiments, stereospecific assignments may result from NOESY connectivities to spatially adjacent groups of protons (Senn et al., 1984; Zuiderweg et al., 1985). Obtaining stereospecific resonance assignments from reference to spatially close spins should eventually be part of the refinement of spatial protein structures determined by NMR.

8.4. COMBINED USE OF SEQUENTIAL ASSIGNMENTS AND ISOTOPE LABELING

The discussion on resonance assignments by isotope enrichment in Section 4.2 showed that labeling of an amino acid type in proteins is quite straightforward, whereas more selective labeling is usually more laborious. While isotope labeling of residue types may be exploited for a variety of experiments (Griffey et al., 1985), combined use with sequential resonance assignments is particularly attractive. Quite generally, assignments should thus be attained for more complex systems, including bigger proteins. Labeling of residue types with ^{13}C, ^{15}N, or other suitable isotopes introduces easily recognizable reference points in known sequence locations. For complex systems one could thus envisage to obtain resonance assignments for peptide segments bounded by such isotope markers, without attempting to work on the entire polypeptide chain. Such site-directed resonance assignments might be used for studies of the active centers in enzymes, or other polypeptide regions of special interest. Alternatively, one might construct mixed polypeptide chains consisting of a combination of perdeuterated and protonated segments. For example, in a protein that is too large or too complex for obtaining complete ^1H NMR assignments, only limited polypeptide segments would then be observed. (Even in H_2O solution, perdeuterated peptide segments would produce no COSY cross peaks, since there are no coupling partners for the amide protons. In NOESY, one would expect exclusively cross peaks with amide protons from the deuterated segments.)

As an illustration, the following describes the use of isotope substitution in the resonance assignments of metallothionein-2 (Neuhaus et al., 1984a). This small protein of 62 residues contains 20 cysteines and a total of 33 amino acids with αCH–βCH_2 side chain spin systems. Identification of the 20 cysteine AMX spin systems was therefore a decisive step. It was based on the knowledge that the protein contains seven metal ions that are all bound to cysteine sulfur atoms (Armitage and Otvos, 1982; see also Fig. 5.40). Homonuclear ^1H COSY spectra were recorded with two metallothionein samples that had been reconstituted with NMR inactive ($I = 0$) $^{112}Cd^{2+}$ or NMR active ($I = \frac{1}{2}$) $^{113}Cd^{2+}$, respectively. ^{112}Cd is not manifested in the COSY cross peak fine structures, whereas the presence of ^{113}C in structures

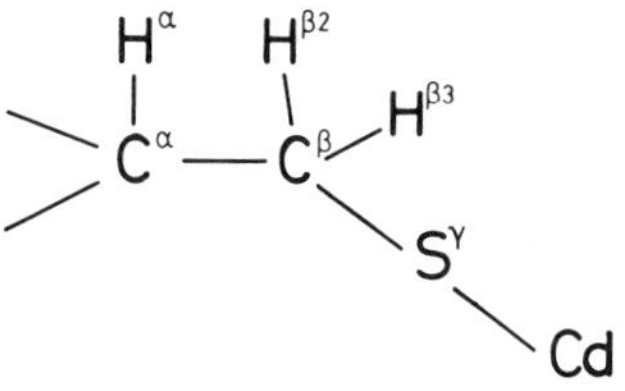

Figure 8.20. Structure of Cys bound to cadmium in metallothionein-2, where Cd is either NMR inactive ^{112}Cd (I = 0) or NMR observable ^{113}Cd (I = $\frac{1}{2}$).

such as that shown in Figure 8.20 results in heteronuclear scalar spin–spin couplings with the β protons. The corresponding splittings are superimposed on the ^{1}H–^{1}H couplings of the Cys AMX spin systems (Fig. 5.28), as is shown in the schemes of Figure 8.21 for a cysteine bound to one or two ^{113}Cd ions. For the αH–βH cross peaks this amounts to having one or two in-phase splittings along ω_2, which displace the complete ^{1}H–^{1}H fine structure pattern. Depending on the relative size of the ^{1}H–^{1}H and ^{1}H–^{113}Cd couplings, very different cross peak fine structures may result, including that some lines can be cancelled because of the antiphase character of the active ^{1}H–^{1}H coupling (Fig. 8.21). In the experimental ^{1}H COSY cross peaks, the coupling to ^{113}Cd can readily be observed (Fig. 8.22A) and the Cys spin systems thus identified, since all noncysteine resonances are unaffected when ^{112}Cd is substituted by ^{113}Cd.

A special feature in the βH–β'H cross peaks is that one observes only two of the four components expected to arise from ^{1}H–^{113}Cd coupling along both ω_1 and ω_2 (Fig. 8.21). The origin of this is that, unlike the protons, the ^{113}Cd nuclei experience no mixing pulse during the homonuclear ^{1}H COSY pulse sequence. The two missing components correspond to situations in which βH would see the ^{113}Cd spin in one orientation at the same time as β'H sees it

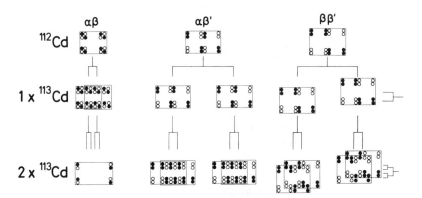

Figure 8.21. Schematic representation of the influence of heteronuclear ^{1}H–^{113}Cd J couplings on the cross peaks of cysteine spin systems in phase sensitive ^{1}H COSY. Filled and empty circles represent positive and negative peaks, respectively (from Neuhaus et al., 1984a).

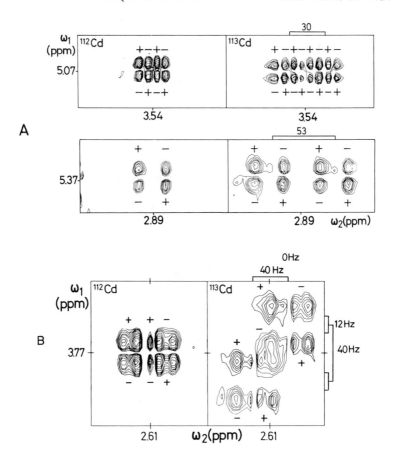

Figure 8.22. Comparison of ¹H COSY cross peaks of Cys in ¹¹²Cd₇ metallothionein-2 and ¹¹³Cd₇ metallothionein-2. *(A) αH–βH* cross peaks of Cys bound to one Cd ion. The size of the ¹H–¹¹³Cd coupling is indicated at the top. *(B) βH–β'H* cross peak of a Cys bound to two Cd ions. The size of the ¹H–¹¹³Cd couplings is indicated at the top and on the right. In (A) and (B) positive and negative peaks are indicated by the appropriate signs (from Neuhaus et al., 1984a).

in the opposite one, which is impossible in this experiment. Figure 8.22*B* shows such an incomplete pattern, which was observed for the *β*H–*β'*H cross peak of a Cys bound to two ¹¹³Cd ions.

8.5. COMPLETE ¹H NOESY CROSS PEAK ASSIGNMENTS

Upon completion of the sequence-specific resonance assignments, all COSY cross peaks are usually identified (Fig. 5.8*C*). In contrast, only the NOESY cross peaks manifesting sequential connectivities and certain intraresidue connectivities were used for obtaining the assignments. Therefore, all NOESY cross peaks in the spectral regions d–f (Fig. 6.8) and numerous cross peaks in the regions a–c (for an illustration, see Fig. 7.14) remain unassigned. The initial step toward determination of the protein conformation then consists of assigning all NOESY cross peaks with the use of the chemical shifts obtained with the sequence-specific resonance assignments (Table 8.3). This laborious procedure should soon be automated, possibly with interactive *peak picking* routines using computer graphics. Ambiguities in the cross peak assignments tend to arise from overlap in crowded regions, in particular for NOE's connecting different aliphatic amino acid side chains (Fig. 6.8, regions e and f). The ambiguities can often be resolved only after the initial determination of a low-resolution 3D structure (Chapters 9 and 10). It must also be emphasized that the extent to which NOESY cross peak assignments can be obtained depends critically on *complete* identification of *all* amino acid spin systems, since in the presence of residual unassigned proton resonances certain cross peak assignments may remain ambiguous.

CHAPTER 9
Polypeptide Secondary Structures in Proteins by NMR

Once sequence-specific resonance assignments have been obtained, further NMR measurements, such as NOE's, spin–spin coupling constants, or amide-proton exchange rates can be attributed to specified locations in the sequence. The occurrence of certain patterns of NMR parameters along the polypeptide chain is then indicative of particular secondary structures. The information thus obtained may be used as a starting point for determination of the tertiary structure, or as a check on spatial structures obtained independently, for example, from distance geometry analysis of the NMR data (Chapter 10). Most important, however, the procedures in this Chapter present an efficient, direct way for secondary structure determination based on qualitative analysis of NOESY spectra, which might also be of interest for applications in industrial product control. Using these techniques, NMR yields more details of the secondary structure than can be obtained with other methods presently available for solution studies. For example, a decisive advantage of the NMR approach when compared to infrared, Raman, or circular dichroism spectroscopy is that the secondary structure elements can be located in the amino acid sequence, and that the number of amino acid residues contained in distinct secondary structure elements can be determined.

9.1. REGULAR SECONDARY STRUCTURES FROM OBSERVATION OF SHORT ¹H–¹H DISTANCES

In Chapter 7 we learned that in addition to the sequential ¹H–¹H distances used for resonance assignments (Chapter 8), the common secondary structures contain medium-range and long-range ^{1}H–^{1}H distances that are sufficiently short for observation by NOE's. Table 7.1 lists distances that would be suitable for distinguishing between different standard secondary structures. The question now arises of how reliable secondary structure identification by qualitative delineation of these short distances would be, since it is well known that globular proteins usually contain somewhat distorted secondary structure elements (Richardson, 1981; Schulz and Schirmer, 1979).

The quantities of interest are the *extent* and *uniqueness* to which a secondary structure element in a globular protein can be identified by observation of a specified short ^{1}H–^{1}H distance. Estimates for these quantities were obtained from a statistical analysis using a group of 19 high-resolution protein crystal structures comprising approximately 3200 residues (Wüthrich et al., 1984a). In these the sequence locations of helical and β-sheet secondary structures were identified using an algorithm by Kabsch and Sander (1983). A first computer search then determined how often a particular distance, for example, $d_{\alpha N}(i,i + 3)$, occurred in the helices, in β sheets, and in the other regions of the polypeptide chains. A second computer search explored how often this distance is shorter than a predetermined limit, for example, $d_{\alpha N}(i,i + 3) \leq 3.5$ Å, in the different secondary structures. The percentage of all the distances $d_{\alpha N}(i,i + 3)$ defined in helical regions that is ≤ 3.5 Å then corresponds to the *extent* to which helical secondary structure is identified by this distance parameter. The *uniqueness* of helix identification by the same parameter is given as the percentage of all distance values $d_{\alpha N}(i,i + 3) \leq 3.5$ Å in the entire protein that is located in helical regions. Corresponding definitions apply for the other secondary structures.

Table 9.1 presents a survey of the extent and uniqueness of helix- and β-

TABLE 9.1. ^{1}H–^{1}H Distances for Regular Polypeptide Secondary Structure Identification.[a]

Secondary Structure	^{1}H–^{1}H Distance constraint (Å)	Extent of Identification	Uniqueness of Identification
αhelix+3$_{10}$helix	$d_{NN} \leq 3.6$	98%	51%
	$d_{\alpha N}(i,i+3) \leq 3.6$	52	81
	$d_{\alpha N}(i,i+3) \leq 4.4$	94	74
	$d_{\alpha N}(i,i+4) \leq 4.4$	65	84
	$d_{NN}(i,i+2) \leq 4.4$	72	62
	$d_{\alpha\beta}(i,i+3) \leq 3.6$	74	79
	$d_{\alpha\beta}(i,i+3) \leq 4.4$	87	71
$\beta + \beta_P$	$d_{\alpha N} \leq 3.0$	98	41

[a] These distances represent an acceptable compromise for high extent and high uniqueness of secondary structure identification (from Wüthrich et al., 1984a).

sheet identification by a selection of distance parameters that were found to be most useful for this purpose. It shows, for example, that 98% of all helical residues are identified in a search with $d_{NN} \leq 3.6$ Å, but the uniqueness of helix identification by this parameter is low, since 49% of all distances $d_{NN} \leq 3.6$ Å were observed in nonhelical peptide segments. Similar numbers were found for the identification of β sheets by the sequential distance $d_{\alpha N} \leq 3.0$ Å. More favorable values for the uniqueness of helix identification are obtained with the six medium-range distances in Table 9.1, but some of these recognize helical segments only to a relatively low extent.

The origin of the low uniqueness of secondary structure identification by the sequential distances $d_{\alpha N}$ and d_{NN} (Table 9.1) lies in the fact that one of these distances is short in *all* local conformations corresponding to one of the allowed regions in the Ramachandran ϕ versus ψ plot (Figs. 7.6 and 7.7). Since regular secondary structures are characterized by the prevalence of identical ϕ and ψ values for all constituent residues, the situation is improved in a search for segments of several successive short sequential distances. Table 9.2 shows that the best results for helix identification using d_{NN} are achieved with five to seven successive distances $d_{NN} \leq 3.6$ Å, which gives an extent of nearly 90% and a uniqueness of 80%.

For β structures the use of intrastrand distances is limited to $d_{\alpha N}$, since in extended polypeptide chains all medium range $^1H-^1H$ distances are too long to be observed by NOE's. The uniqueness of β sheet identification by segments of successive short sequential distances $d_{\alpha N}$ (Table 9.3) is markedly lower than those for helix identification by d_{NN}. In part this arises because globular proteins contain a sizable number of residues in extended peptide segments that are not part of a β sheet. Further distinction of those extended strands that *are* located in β sheets can then be obtained from observation of short long-range $^1H-^1H$ distances between neighboring strands (Section 7.4, Fig. 7.13). In principle, observation of two such distances would be sufficient to identify two neighboring strands in a β sheet and to determine the relative polarity of the two strands (Fig. 7.13). Statistical studies with pro-

TABLE 9.2. Extent and Uniqueness of Helix Identification by Segments of Several Subsequent Short Sequential Distances d_{NN}.[a]

Length of segment	$d_{NN} \leq 3.0$ Å		$d_{NN} \leq 3.6$ Å	
	Extent	Uniqueness	Extent	Uniqueness
1	86	54	98	51
3	80	74	97	68
5	68	83	91	78
7	55	85	85	80

[a] From Wüthrich et al. (1984a).

TABLE 9.3. Extent and Uniqueness of β-Sheet Identification by Segments of Several Subsequent Short Sequential Distances $d_{\alpha N}$.[a]

Length of segment	$d_{\alpha N} \leqslant 2.6$ Å		$d_{\alpha N} \leqslant 3.0$ Å	
	Extent	Uniqueness	Extent	Uniqueness
1	95	46	98	41
3	90	55	95	48
5	79	63	88	53
7	59	65	74	57

[a] From Wüthrich et al. (1984a).

tein crystal structures showed that with the typical length of the β sheets in globular proteins, one can expect to observe six to eight short distances between any two neighboring β strands, which is usually sufficient to further pinpoint irregular β-sheet features, such as β bulges (Richardson, 1981).

Overall, the identification of helical and β sheet regular secondary structures, which both contain a dense network of short $^1H-^1H$ distances (Figs. 7.11 and 7.13), has to rely on a combined observation of all suitable distance parameters (Tables 7.1 and 9.1–9.3). While there is no single distance constraint that would lead to extensive and unique identification of a particular structure type, it is unlikely that either a helix or a β sheet would escape detection when the search extends over all these parameters. In practice, the information on the sequential distances $d_{\alpha N}$ and d_{NN} gathered during sequential assignments (Fig. 8.18) already provides a strong indication of the sequence locations of helices and β strands (Tables 9.2 and 9.3). These can subsequently be confirmed or discarded using the medium-range $^1H-^1H$ distances for helices and the interstrand long-range distances for β sheets. With the exception of $d_{\alpha\beta}(i,i + 3)$ in helices and $d_{\alpha\alpha}(i,j)$ in antiparallel β sheets, the distances of interest will all be manifested in the regions a and b of NOESY spectra recorded in H_2O (Fig. 6.8), which contain only few unassigned cross peaks after the identification of the intraresidue and sequential connectivities (Fig. 7.14). $d_{\alpha\alpha}(i,j)$ and $d_{\alpha\beta}(i,i + 3)$ are observed in the regions d and e, respectively, of NOESY spectra recorded in D_2O (Fig. 6.8). Generally, β sheets are easiest to characterize in detail. For helical structures one needs to observe only a small proportion of all potentially available short distances (Tables 7.1 and 9.1) to characterize the main body of the helix, whereas accurate identification of the two ends of a helix can be more difficult and may require additional data (Sections 9.2 and 9.3).

The procedures discussed so far in this section allow one to distinguish between peptide segments forming helical structures, parallel β sheets, antiparallel β sheets, and all others. Table 7.1 further indicates that measurements of short $^1H-^1H$ distances might permit discrimination between α helices

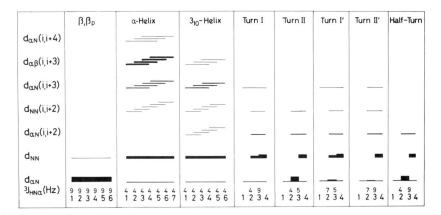

Figure 9.1. Survey of the sequential and medium range $^1H-^1H$ NOE's and the spin–spin coupling constants $^3J_{HN\alpha}$ in the following secondary structures: parallel or antiparallel β sheet, α helix, 3_{10} helix, turns of types I, II, I', and II', and a half-turn derived from a type II tight turn (see text). The numbers at the bottom represent the amino acid residues in the secondary structure elements and the values of $^3J_{HN\alpha}$. Short $^1H-^1H$ distances are indicated by lines linking the residues that contain the connected hydrogen atoms; the thickness of the lines is proportional to r^{-6} and thus represents the NOESY cross-peak intensities (from Wagner et al., 1986a).

and 3_{10} helices. The NOESY cross peak intensities that might be used for this distinction are visualized in Figure 9.1, which was constructed from the data of Table 7.1 with the assumption that the structures are rigid and that the same correlation function, $f(\tau_c)$ in Eq. (6.1), applies for all connectivities. The figure shows that most NOE's are identical for the two helices and that a distinction between α helix and 3_{10} helix has to rely primarily on observing $d_{\alpha N}(i, i + 4)$ in the former and $d_{\alpha N}(i, i + 2)$ in the latter. It should also be noted that both of these distances are expected to yield very weak NOE's, so that it is unlikely that a complete set over the whole length of the helix will be seen for either one. Support for the prevalence of an α helix may further be obtained from observation of rather strong NOE's $d_{\alpha\beta}(i, i + 3)$.

9.2. REGULAR SECONDARY STRUCTURES FROM SPIN–SPIN COUPLINGS $^3J_{HN\alpha}$

The use of vicinal spin–spin couplings for studies of molecular conformations relies on a general relation between the size of the spin–spin coupling constant ^3J and the intervening torsion angle θ (Karplus, 1959). The dependence of $^3J_{HN\alpha}$ on the torsion angle ϕ in polypeptides was surveyed by Bystrov (1976) and more recently (Pardi et al., 1984) Eq. (9.1) was derived from correlating the ϕ angles in the crystal structure of BPTI and the $^3J_{HN\alpha}$

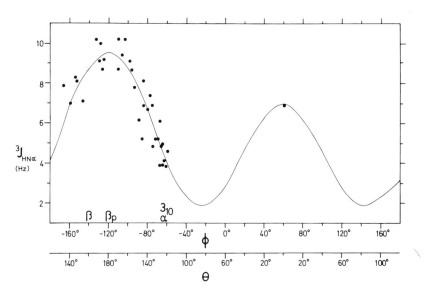

Figure 9.2. Plot of $^3J_{HN\alpha}$ versus the torsion angle $\theta = |\phi - 60°|$ in BPTI. ϕ was obtained from the crystal structure and $^3J_{HN\alpha}$ from solution studies. The filled circles represent individual amino acid residues. The curve corresponds to the best fit obtained with Eq. (9.1). ϕ values for regular secondary structures are indicated by α, 3_{10}, β, and β_p (from Pardi et al., 1984).

values measured in solution (Fig. 9.2), with $\theta = |\phi - 60°|$ and $^3J_{HN\alpha}$ given in hertz.

$$^3J_{HN\alpha} = 6.4 \cos^2 \theta - 1.4 \cos \theta + 1.9 \tag{9.1}$$

With the use of Eq. (9.1), measurements of $^3J_{HN\alpha}$ (Fig. 5.29) present a valuable complementation of NOE distance data for studies of the polypeptide-backbone conformation in proteins.

When vicinal spin–spin coupling constants are correlated with the intervening torsion angles via a Karplus-type relation, one is left, in general, with some ambiguities, since as many as four different values for the torsion angle may correlate with a predetermined coupling constant (Fig. 9.2). In proteins, however, the dihedral angles ϕ for all amino acid residues except glycine are concentrated in the range of -30 to $-180°$ (Richardson, 1981) (the BPTI data in Fig. 9.2 include only a single exception to this). Using Eq. (9.1) the following values for $^3J_{HN\alpha}$ in regular secondary structures are obtained:

$$\alpha \text{ helix } (\phi = -57°), \ ^3J_{HN\alpha} = 3.9 \text{ Hz.}$$
$$3_{10} \text{ helix } (\phi = -60°), \ ^3J_{HN\alpha} = 4.2 \text{ Hz.}$$
$$\text{Antiparallel } \beta \text{ sheet } (\phi = -139°), \ ^3J_{HN\alpha} = 8.9 \text{ Hz.}$$
$$\text{Parallel } \beta \text{ sheet } (\phi = -119°), \ ^3J_{HN\alpha} = 9.7 \text{ Hz.}$$

TABLE 9.4. Extent and Uniqueness of Secondary Structure Identification Using Spin–Spin Couplings $^3J_{HN\alpha}$.[a]

Length of segment	Extent	Uniqueness
Helix, $^3J_{HN\alpha} < 6.0$ Hz		
1	76	54
3	54	83
5	41	87
β-sheet, $^3J_{HN\alpha} > 7.0$ Hz		
1	87	44
3	72	54
5	62	48

[a] Identification of helical secondary structure by segments of several subsequent spin–spin couplings $^3J_{HN\alpha} < 6.0$ Hz, and of β-sheet structure by several subsequent couplings $^3J_{HN\alpha} > 7.0$ Hz (from Pardi et al., 1984).

These numbers provide a basis for identification of regular secondary structures from measurements of $^3J_{HN\alpha}$. A statistical study on a group of protein crystal structures conducted along the same lines as for the ^1H–^1H distances (Table 9.1) showed that a similar situation prevails as for the sequential distances d_{NN} (Table 9.2) and $d_{\alpha N}$ (Table 9.2). Individual small or large couplings $^3J_{HN\alpha}$ identify helical or β-sheet structures, respectively, to a high extent but with low uniqueness (Table 9.4). For helices, a uniqueness of over 80% is attained with segments of three to five subsequent residues with $^3J_{HN\alpha} < 6.0$ Hz. For β sheets the corresponding uniqueness is lower, again primarily because globular proteins contain extended peptide segments also outside of β sheets.

Overall, the spin–spin coupling constants $^3J_{HN\alpha}$ present supporting evidence for helical and β-sheet regular secondary structures identified by short ^1H–^1H distances (Section 9.1). Furthermore, observation of segments of three or more sequentially neighboring residues with small coupling constants is a reliable, independent criterion for identification of helical regions in polypeptide chains (Table 9.4).

9.3. HYDROGEN BONDS AND AMIDE-PROTON EXCHANGE

The formation of hydrogen bonds between backbone amide protons and backbone carbonyl oxygens is an outstanding trait of the common polypeptide secondary structures (Richardson, 1981; Schulz and Schirmer, 1979). In the α helix and the 3_{10} helix one has the bonds CO_i—NH_{i+4} and CO_i—HN_{i+3}, respectively, and in β sheets one observes a dense network of hydrogen bonds between neighboring peptide strands (Fig. 7.13). In the regular secondary structures all amide protons are involved in hydrogen bonds, with the following exceptions: The first four residues in an α helix; the first three residues in a 3_{10} helix; every second residue in the peripheral strands of β sheets.

Experience has shown that the individual amide-proton exchange rates measured with the experiments of Figures 3.13 or 5.35 provide a surprisingly faithful manifestation of hydrogen bonding in regular secondary structures. When combined with sequence-specific resonance assignments, the exchange data can be plotted versus the amino acid sequence. For BPTI (Fig. 1.3) such a plot contains two segments, residues 15 to 20 and 29 to 35, in which slowly and rapidly exchanging amide protons alternate along the chain. These two segments form peripheral strands of the antiparallel β sheet in BPTI, where the amide protons are alternatively hydrogen bonded or exposed (Fig. 7.13), and slow exchange was measured exclusively for those groups that are hydrogen bonded in the X-ray structure. In the segment from residues 20 to 25, which forms the central strand of the β sheet and has therefore all amide protons in hydrogen bonds, very slow exchange was measured for all residues. A second continuous segment of slow exchange rates from residues 51 to 55 corresponds to an α helix that extends from residues 47 to 55.

Overall, individual amide-proton exchange rates can provide fully independent, supporting evidence for regular secondary structures identified by NOE observation of short $^1H-^1H$ distances, or by $^3J_{HN\alpha}$ spin–spin couplings. They present particularly clear data for the identification of the C-terminal end of α or 3_{10} helices, and for the distinction between peripheral and nonperipheral peptide strands in β sheets.

9.4. IDENTIFICATION OF TIGHT TURNS

The definition of a *tight turn* (Richardson, 1981) involves a peptide segment of *four residues*, as exemplified by Figure 7.12. However, since for residues 1 and 4 only the orientation of the carbonyl group, or the NH group, respectively, is defined, tight turns or *hairpin loops* may alternatively be considered as *two-residue loops*, with the residues 2 and 3 forming the actual turn (Sibanda and Thornton, 1985). The conformation of these loops is determined by the four torsion angles ϕ_2, ψ_2, ϕ_3, and ψ_3, and it was suggested that all tight turns fall into one of four major classes (Richardson, 1981): *Type I* and *type II* turns as defined in Figure 7.12, and corresponding *mirror-image* versions *type I′* and *type II′*, where the torsion angles have opposite signs. The NOE patterns and spin–spin couplings $^3J_{HN\alpha}$ expected for these structures are shown in Figure 9.1.

There are several factors that make the identification of tight turns by NOE distance constraints, $^3J_{HN\alpha}$, and amide-proton exchange studies less reliable than for regular secondary structures. First, Figure 9.1 shows that the most outstanding features in tight turns are isolated short distances $d_{\alpha N}$ and d_{NN}, and isolated extreme values of $^3J_{HN\alpha}$. For the reasons discussed in connection with the regular secondary structures (Tables 9.2–9.4), the uniqueness of turn identification by these parameters is low. Second, the distance constraints in tight turns are similar to those in helical peptide segments. For turns of type I and I′ both the sequential and medium-range $^1H-^1H$ distances coincide closely with those in the 3_{10} helix, and for turns II and II′ qualitative differences from the 3_{10} helix prevail only for the sequential distances $d_{\alpha N}$ and d_{NN} between residues 2 and 3. Third, probably because tight turns in globular proteins tend to be exposed near the molecular surface (Richardson, 1981; Schulz and Schirmer, 1979), the exchange of the hydrogen-bonded amide protons in these structure elements (Fig. 7.12) is generally too rapid to provide evidence for the identification of the hydrogen bonds. A priori, this all limits the identification of tight turns with the empirical pattern recognition approach described in this chapter to certain favorable situations. For example, it is straightforward to locate hairpin tight turns that connect neighboring strands in antiparallel β sheets. In contrast, turns located at either end of an α helix or 3_{10} helix are difficult to identify without additional knowledge on the overall three-dimensional arrangement of the polypeptide chain. Overall, with the exception of the hairpin turns in antiparallel β sheets, it is therefore advisable to defer locating tight turns to the tertiary structure determination (Chapter 10).

Once the sequence locations of the tight turns have been identified, there remains the question of to what extent different turns can be distinguished by the qualitative criteria of Figure 9.1. The figure implies that it should be possible to distinguish between type I and type II families of turns. However, as long as the experimental NOE data consist of qualitative distance constraints rather than quantitative distance measurements, the parameters of Figure 9.1 are not sufficient to define *classical* tight turn types (Richardson, 1981), but are compatible with a range of turn conformations resembling the classical types. This is best illustrated with an example. In type II turns a strong sequential NOE $d_{\alpha N}$ between residues 2 and 3 and a NOE-observable short distance $d_{\alpha N}(2,4)$ constrain ψ_2 and ψ_3 to values near $+120$ and $0°$, respectively. The torsion angle ϕ_3, however, is hardly constrained by the $^1H-^1H$ NOE's, since there are only small variations in the distances $d_{NN}(3,4)$ and $d_{\alpha N}(2,4)$ if ϕ_3 is rotated from $+90°$ through 0 to $-90°$ (Fig. 7.12), resulting in a *half-turn* structure (Fig. 9.1). While in this case the NOE data are compatible with a wide range of variations of the turn structure, the spin–spin coupling constant $^3J_{HN\alpha}$ for residue 3 can be used to further constrain the conformation, since it adopts values of 5Hz for $\phi = 90°$ in type II turns, and 9 Hz for $\phi = -90°$ in half-turns (Wagner et al., 1986a).

9.5. APPLICATIONS

A first example is BUSI, a protein with 57 amino acid residues (Williamson et al., 1984). Figure 9.3 presents a survey of the experimental data used for the secondary structure determination. Strong sequential NOE's $d_{\alpha N}$ indicated that the peptide segments 4–7, 14–26, 29–33, and 49–55 are in an extended form. For a large proportion of these residues this conclusion was corroborated by large spin–spin couplings $^3J_{HN\alpha}$. NOE's corresponding to long-range backbone distances then showed that the segments 23–27, 29–32, and 50–55 form a triple-stranded β sheet, with the hydrogen bonds identified at the bottom of Figure 9.3 and with residues 52 and 53 forming a β bulge. The residues 23–27 in the central strand all have slowly exchanging amide protons. In the qualitative distinction between slow and rapid exchange used for Figure 9.3, most NH exchange rates in the peripheral β strands are classified as measurably slow (which essentially means that they could be observed in D_2O solution). In a plot of quantitative NH exchange rates versus the sequence, however, the typical pattern for peripheral β strands (Fig. 1.3) was observed for both segments 29–32 and 50–55, with an irregu-

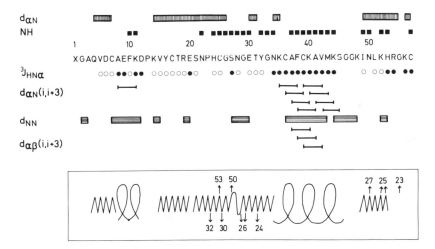

Figure 9.3. Amino acid sequence of BUSI (X = pyroglutamic acid) and experimental data used for secondary structure identification. The following symbols are used: $d_{\alpha N}$ and d_{NN}, a shaded bar indicates that a strong NOE was observed. NH, a filled square indicates slow NH exchange. $^3J_{HN\alpha}$, open circles indicate > 7.0 Hz, filled circles < 6.0 Hz. $d_{\alpha N}(i,i+3)$ and $d_{\alpha\beta}(i,i+3)$, a line connects the two residues between which a NOE was observed. The box at the bottom describes the secondary structure. A zig-zag line indicates extended chain, helices are drawn as such. For those extended segments that are part of a β sheet, arrows indicate the residues where the amide proton is involved in hydrogen bonds across the β sheet, and the numbers identify the sequence locations of the carbonyl oxygens in these hydrogen bonds (from Williamson et al., 1984).

larity at the β bulge in positions 52 and 53 (Wüthrich et al., 1984b). Indications for helical structure were initially obtained from observation of strong NOE's d_{NN} in the segments 7–12, 27–30, 36–43, and 44–48. For the first of these segments an additional short distance $d_{\alpha N}(i,i + 3)$ between residues 8 and 11, slow NH exchange for residues 10 and 11, and small couplings $^3J_{HN\alpha}$ for four of the five residues provided additional support for the presence of a short helix. The segment 27–30 is in the region of the hairpin turn connecting the first two β strands. The third segment covers a region in the amino acid sequence where medium range NOE's $d_{\alpha N}(i,i + 3)$ and $d_{\alpha\beta}(i,i + 3)$, small couplings $^3J_{HN\alpha}$ and slow NH exchange clearly showed that there is a quite regular α helix extending from residues 34 to approximately 45. The segment of short sequential distances d_{NN} from residues 44–48 falls partly into this same helix and possibly covers a type I tight turn that would connect the helix with the following secondary structure element. A schematic presentation of these data is afforded at the bottom of Figure 9.3, which indicates the helical regions and the extended peptide segments, and describes the hydrogen bonding network for those extended segments that form a β sheet.

A *diagonal plot* of the NOE data (Fig. 9.4) affords a compact presentation

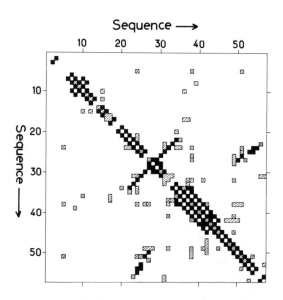

Figure 9.4. Diagonal plot of the NOE's observed in BUSI. Both axes represent the amino acid sequence. A filled square at the position (x,y) indicates that a NOE between backbone protons (NH or αH) of the two residues in the sequence locations x and y was observed. A shaded square indicates a NOE between NH or αH of one residue and a side chain proton in another residue, and a square containing a cross indicates a NOE between side chain protons of two different residues. Where two residues are connected by more than one NOE, only the one that involves the largest number of backbone protons is shown (from Williamson et al., 1985).

of the distance constraints measured in a protein, which is particularly help-ful for the secondary structure determination. The manifestations of differ-ent polypeptide conformations are closely similar to those in the diagonal plots of $\alpha C-\alpha C$ distances used for presentation of protein crystal structures (Richardson, 1981; Schulz and Schirmer, 1979). Since with the exception of $d_{\alpha\beta}(i,i+3)$ the secondary structure determination relies on NOE's between backbone atoms, only the filled squares in Figure 9.4 need to be considered. In BUSI the sequential NOE's appear as a nearly complete line of squares immediately adjacent to the diagonal. The helical regions are characterized by lines of squares running parallel to the diagonal at a distance of three sequence positions, with some additional NOE's at two or four positions from the diagonal. The regular antiparallel β sheet formed by the strands 23–27 and 29–32 is manifested by a continuous line of squares oriented perpen-dicular to the diagonal. For the antiparallel combination of the strands 23–27 and 50–55 this perpendicular array of NOE's can also be recognized, but it is somewhat distorted because of the β bulge formed by residues 52 and 53. A parallel β sheet would produce a line of squares parallel to the diagonal at a distance determined by the relative sequence locations of the two neighbor-ing strands. Quite generally, a diagonal plot presentation of the NOE data can provide a clear visualization of the characteristic patterns of short ^{1}H–^{1}H distances for the common polypeptide secondary structures (Table 7.1, Fig. 9.1).

A second example is Tendamistat, which is an α-amylase inhibitor con-sisting of a polypeptide chain with 74 amino acids (Kline and Wüthrich, 1985). Figure 9.5 presents a survey of sequential NOE's, spin–spin cou-plings $^3J_{HN\alpha}$, and NH exchange rates. The sequential NOE's analyzed for the

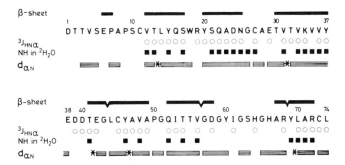

Figure 9.5. Sequence of Tendamistat and NMR data used for the characterization of the secondary structure. Below the sequence, open circles indicate $^3J_{HN\alpha} \gtrsim 8.0$ Hz, filled squares indicate slow NH exchange, and a shaded bar indicates that a strong sequential NOE $d_{\alpha N}$ was observed. Stars indicate locations for which $d_{\alpha N}$ could not be studied due to overlap of the αH position with the decoupler frequency. Above the sequence, solid lines indicate the sequence positions of seven β strands, with nicks identifying the locations of three β bulges (from Kline and Wüthrich, 1985).

A

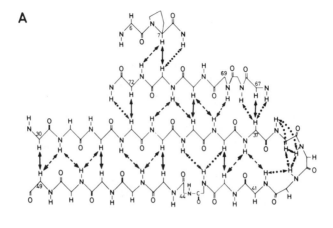

B

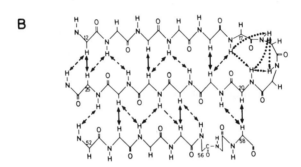

Figure 9.6. Antiparallel β structures in Tendamistat. Interstrand ^{1}H–^{1}H NOE's are indicated by arrows: Solid arrows, $d_{\alpha\alpha}(i,j)$; broken arrows, $d_{\alpha N}(i,j)$ and $d_{NN}(i,j)$ observed in D_2O solution; dotted arrows, $d_{\alpha N}(i,j)$ and $d_{NN}(i,j)$ observed only in H_2O solution (from Kline and Wüthrich, 1985).

resonance assignments gave a strong indication that Tendamistat is a β protein. Extensive extended regions were also implicated by the large $^3J_{HN\alpha}$ couplings. Observation of numerous long-range NOE's then showed that these extended peptide segments are combined into two antiparallel β sheets (Fig. 9.6). Both sheets contain three major strands, and there is a short fourth strand attached to one of these structures (Fig. 9.6A). It is readily apparent from Figure 9.6 that the long-range NOE's $d_{\alpha\alpha}(i,j)$ (Fig. 9.7) have a pivotal role in the NMR characterization of antiparallel β structures. In all, a dense network of interstrand NOE's was observed (Fig. 9.6), and therefore the locations of three β bulges and details on the nature of the turns connecting the β strands could also be determined. Finally, the NH exchange rates clearly supported the fact that the peptide segments 21–26, 33–37, and 69–72

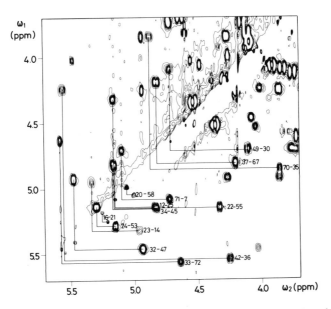

Figure 9.7. Region of the ¹H NOESY spectrum of Tendamistat containing the αH–αH cross peaks corresponding to the solid arrows in Figure 9.6 (0.007 M, D_2O, pD 3.2, 50°C; 500 MHz; τ_m 200 ms; absorption mode, diagonal suppression; digital resolution 4.5 Hz/point in ω_1, 2.2 Hz/point in ω_2). Corresponding cross peaks above and below the diagonal are connected by lines and identified by the sequence locations of the connected residues (from Kline and Wüthrich, 1985).

form nonperipheral strands, whereas the segments 13–17, 30–33, 41–49, and 52–57 display exchange patterns of the kind expected for peripheral β strands (Fig. 9.5).

CHAPTER 10
Three-Dimensional Protein Structures by NMR

The selection of the methods for 3D structure determination is primarily dictated by the data obtained from the NMR measurements. Since these are fundamentally different from X-ray crystal-diffraction data, for which highly efficient techniques of protein structure determination and refinement are available, new avenues must be developed and explored. So far the groundwork has been laid for several different approaches. One of these follows the hierarchical ladder of protein conformations (Schulz and Schirmer, 1979). In a first step, the secondary structure is identified as described in Chapter 9. A three-dimensional molecular model is then constructed with the use of additional, long-range NMR constraints to determine the relative spatial locations of the secondary structure elements. A second approach uses a mathematical procedure called *embedding* (Havel et al., 1983; Havel and Wüthrich, 1984; Sippl and Scheraga, 1985), which produces spatial polypeptide structures that are an approximate fit to the ensemble of all conformational constraints imposed by the NMR data and by the covalent structure. This approximate fit is then further improved by numerical optimization. Both procedures can be supplemented with structure refinement routines, for example, using classical techniques for energy minimization or methods of molecular dynamics calculations. A third approach uses restrained molecular dynamics calculations. In the following these methods for structure determination are introduced, mainly by way of describing practical applications. Independent of the choice of technique for the structural analysis, the outcome of a structure determination will always depend primarily on the correct NMR assignments (Chapter 8) and the quality of the additional NMR data. Therefore, discussion of correlations between NMR data and protein conformation is an important part of this chapter.

10.1.   NMR INPUT FOR PROTEIN STRUCTURE DETERMINATION

The NMR input for structure determination consists primarily of the $^1H-^1H$ distance constraints derived from NOE measurements. For the reasons outlined in Chapter 6, we have so far made only *qualitative* use of NOESY, as is illustrated with the following description of the data for BUSI (Williamson et al., 1985). Similar data handling was previously used with micelle-bound glucagon (Braun et al., 1983) and has more recently been employed with Tendamistat (Kline et al., 1986) and with metallothionein-2 (Braun et al., 1986).

In the NOESY spectra of BUSI recorded in H_2O (Fig. 5.37) and D_2O, a total of 556 cross peaks were identified. The NOE intensities were derived from the peak heights measured either in cross sections or by counting the contour lines in contour plots. Build-up curves were obtained from experiments with mixing times of 50, 80, 100, 120, and 200 ms, and peaks arising from spin diffusion were eliminated. For the remaining cross peaks, the peak heights at 80 ms were taken to represent the relative initial buildup rates. The sequential NOE's were classified as weak, medium, or strong, and the corresponding upper distance constraints were calibrated relative to the standard sequential distances (Table 7.1) in the previously identified helical and β-sheet segments (Table 10.1). Throughout, the lower bounds are equal to the sum of the van der Waals radii of the atoms at close approach. The same calibrations were used for intraresidue NOE's between NH, αH, and βH. NOE's between protons separated by more than four single bonds in the covalent structure were attributed a distance range 2.0–4.0 Å if they involved exclusively backbone NH and αH, or a range 2.0–5.0 Å if they involved side chain protons (Table 10.1). These different upper bounds were selected on the basis of the uniform averaging model (Fig. 6.14), considering that the amino acid side chains are overall more flexible than the backbone. The decision not to interpret the medium-range and long-range NOE's in more detail resulted in a considerable simplification of the analysis of the NOESY spectra. This obviously leaves room for improvements in future studies, where further quantitation of medium- and long-range constraints

TABLE 10.1. NOE Distance Constraints Used to Prepare the Input for the 3D Structure Determination of BUSI

NOE Intensity	Intra-residue or Sequential, NH, αH, βH	Medium- or Long-range, NH, αH	Medium- or Long-range, side chain H
weak	2.0 - 4.0	2.0 - 4.0	2.0 - 5.0
medium	2.0 - 3.0	2.0 - 4.0	2.0 - 5.0
strong	2.0 - 2.5	2.0 - 4.0	2.0 - 5.0

can be expected to result in improved structures (Havel and Wüthrich, 1985).

In the interpretation of the NOE's one must also allow for the *lack of stereospecific resonance assignments* (Section 8.3; with the exception of Gly αCH_2, this exclusively affects NOE's with side chain protons). A set of *pseudostructures* for amino acids was defined for this purpose (Wüthrich et al., 1983). Whenever there is a NOE with one or all protons in a group for which no stereospecific assignments were obtained, the corresponding distance constraint is measured relative to a pseudoatom located centrally with respect to these protons. Since the experimental NOE involves the protons rather than the pseudoatom, a correction must be added to the NOE distance constraint. This correction is set equal to the maximum possible error. For interresidue NOE's it therefore corresponds to the distance between the pseudoatom and the protons that it replaces, and for intraresidue NOE's it is equal to the maximum sterically allowed difference between the relevant distances to the protons and to the pseudoatom, respectively. Table 10.2

TABLE 10.2. Amino Acid Pseudostructures for the Initial Interpretation of NOE Distance Constraints in Proteins

Amino acid	Pseudo-structure[a]	intra-residue correction[b]	inter-residue correction[b]
Gly	NH—LA—CO ⋮ P'A		m
Ala	CH | MB	$i(N \rightarrow \beta)$	m
Ile (Thr)	CH | CH PG⋯ ⋰LG ＼MG | MD	$m(N \rightarrow M\gamma, P\gamma)$ $i(\alpha \rightarrow M\gamma, P\gamma)$ $m(\alpha \rightarrow M\delta)$	m m
Val	CH | CH ╱ ⋮ ＼ MG1 QG MG2	$m(N \rightarrow M\gamma)$ $s(N \rightarrow Q\gamma)$ $i(\alpha \rightarrow M\gamma)$ $i(\alpha \rightarrow Q\gamma)$	q, m
Leu	CH | LB⋯PB | CH ╱ ⋮ ＼ MD1 QD MD2	$i(N \rightarrow P\beta)$ $m(\alpha \rightarrow M\delta)$ $s(\alpha \rightarrow Q\delta)$	m q, m

TABLE 10.2. (Continued)

Amino acid	Pseudo-structure[a]	intra-residue correction[b]	inter-residue correction[b]
Lys	C H		
(Ser, Asp, Asn,	|		
Cys, His, Trp, Glu,	L B···PB	$i(N \rightarrow P\beta)$	m
Gln, Met, Arg, Pro)	|		
	L G···PG	$m(N \rightarrow P\gamma)$	m
		$i(\alpha \rightarrow P\gamma)$	
	|		
	L D···PD	$m(\alpha \rightarrow P\delta)$	m
	|		
	L E···PE		m
	|		
	N Z		
Phe	C H		
(Tyr)	|		
	L B···PB	$i(N \rightarrow P\beta)$	m
	|		
	C1	$r(N \rightarrow C1)$	r
	K6 ╱ ╲ K2	$m(\alpha \rightarrow C1)$	
	| QR |	$q(N \rightarrow QR)$	q
	K5 ╲ ╱ K3	$r(\alpha \rightarrow QR)$	
	C4	$m(N \rightarrow C4)$	r
	|	$i(\alpha \rightarrow C4)$	
	H4		

[a] P, Q, and QR are dimensionless pseudoatoms used as reference points for NOE distance constraints. P and Q are centrally located relative to the two protons of CH_2 groups and the two methyl groups in Val and Leu, respectively. QR is in the ring center in Phe and Tyr. K, L, and M are spheres of radii 1.5, 1.6, and 1.8 Å representing, respectively, the volume occupied by ring CH groups in Phe and Tyr, by CH_2 groups, and by CH_3 groups. K and L are at the carbon positions. M is in the center of the three methyl protons at a distance of 0.36 Å from the methyl carbon atom, and it serves also as a reference point for NOE distance constraints to the methyl protons. The indices A, B, G, D, E, and Z, are used in place of the more common Greek letters to identify the side chain atom positions.
[b] $i = 0.6$ Å, $m = 1.0$ Å, $s = 1.7$ Å, $r = 2.0$ Å, $q = 2.4$ Å. Intraresidue corrections are only indicated when they are different from the corresponding interresidue corrections. (From Wüthrich et al., 1983.)

lists the pseudostructures for 7 amino acids, from which the pseudostructures for all 20 common amino acids can be derived, as is also indicated in the Table. In addition to the dimensionless pseudoatoms used as reference points for NOE distance constraints, these pseudostructures contain spherical pseudoatoms representing the van der Waals volumes of CH, CH_2, and CH_3 groups. Through this, the number of atoms in the polypeptide chain is reduced and the computation time for structure determinations can be shortened.

In Table 10.2 the reference point for NOE distance constraints with αCH_2 of Gly is defined by PA, whereas the volume of αCH_2 is represented by LA. In Ala and Ile, M is the reference point for NOE's to the methyl protons *and* represents the volume of the CH_3 groups. The data for Ile γCH_3 are also valid for Thr γCH_3. In Val and Leu, NOE's to the methyl protons are referred to QG and QD, respectively, and the volume of each individual CH_3 group is represented by M. In Lys, each CH_2 group is represented by two pseudoatoms P and L. Identical corrections apply to the corresponding CH_2 groups in Ser, Asp, Asn, Cys, His, Trp, Glu, Gln, Met, Arg, and Pro. For the ring protons of His and Trp, unique assignments can be obtained and no pseudoatoms are used. In the aromatic rings of Phe and Tyr, QR is the reference point for NOE's when the resonances of the 2,6 and 3,5 protons cannot be resolved, otherwise the NOE's are referred to C1 and C4, respectively. K represents the volume of the CH groups 2, 3, 5, and 6 in these rings.

In a presentation of the NOE distance constraints using the format of Table 10.3, the first column lists the constraints on the local backbone conformation, the second column the constraints on the global backbone fold, and the third column the long-range constraints on the amino acid side chains. (Usually a fourth column is added to list the intraresidue constraints on the side chain conformations.) To visualize the distribution of the different types of constraints along the polypeptide chain, each constraint is listed twice, for both residues connected by the NOE. For each constraint the upper distance bound derived from NOESY with the use of Table 10.1 is listed and, where applicable, corrections for the use of pseudoatoms are identified. The following are two examples: The distance constraint of 4.0 Å between the amide protons of Gly-33 and Asn-34 (Table 10.3) corresponds to observation of a weak NOE (Table 10.1) between these protons. The distance constraint of 6.0 Å between the amide proton of Gly-33 and the methyl protons of Ala-37 is based on observation of a weak NOE, corresponding to a constraint of ≤ 5.0 Å, to which a correction of 1.0 Å was added to account for the use of the pseudoatom M in Ala. The complete listing for BUSI contained 202 NOE constraints. Of the total of 556 NOE's identified in the NOESY spectra, 354 intra-residue and sequential NOE's were not included in the input for the structure determination, since with the wide distance ranges (Table 10.1) and the corrections for pseudoatoms (Table 10.2) these NOE's did not represent meaningful constraints. [For example, a constraint in the range 2.0–4.0 Å is meaningless for a distance $d_{\alpha N}$, which cannot be longer than 3.6 Å (Fig. 7.6).]

Conformational constraints can also be derived from spin–spin coupling constants. In BUSI, the ϕ angles corresponding to $^3J_{HN\alpha} \geq 8.0$ Hz were confined to the range -160 to $-80°$, the ϕ angles corresponding to $^3J_{HN\alpha} \leq 5.5$ Hz to the range -90 to $-40°$ [Eq. (9.1)], and the χ^1 angles of Val, Ile, and Thr corresponding to $^3J_{\alpha\beta} \geq 8.0$ Hz were attributed a range of $\pm 30°$ about the trans position of αH relative to βH.

If the locations of backbone amide-proton to carbonyl-oxygen hydrogen bonds in regular secondary structures can be inferred from NOE's and slow amide-proton exchange rates, tighter constraints may be imposed than those in Table 10.1. In BUSI, ranges of 1.8–2.0 Å for the H–O distance, and 2.7–

TABLE 10.3. NOE Distance Constraints for Residues 32–35 in BUSI

Sequential[a]	Medium-range and Long-range backbone[a]	Long-range[a]
Tyr32[b]		
	HN H24 HN 4.0	HN T31 MG 6.0 m
		QR A37 MB 8.4 qm
		PB A37 MB 7.0 mm
		PB F38 QR 8.4 qr
		QR F38 HA 7.4 q
		QR A41 MB 8.4 qm
		QR I49 MG 8.4 qm
Gly33		
HN N34 HN 4.0	PA P23 HA 5.0 m	PA T31 MG 7.0 mm
	HN A37 MB 6.0 m	
Asn34		
HN G33 HN 4.0	HA N22 HN 4.0	ND E20 HN 6.0 m
HA K35 HN 2.5	HA N22 PB 6.0 m	ND E20 PB 7.0 mm
HN K35 HN 4.0	PB N22 HN 6.0 m	HA H24 H2 5.0
PB K35 HN 3.5 m	PB C36 HN 6.0 m	
	HN A37 HN 4.0	
	HN A37 MB 6.0 m	
	HN F38 HN 4.0	
Lys35		
HN N34 HA 2.5	HN N22 PB 6.0 m	HN N22 ND 6.0 m
HN N34 HN 4.0	HA F38 HN 4.0	HA H24 H2 5.0
HN N34 PB 3.5 m	HA F38 PB 6.0 m	HN H24 H2 5.0
HN C36 HN 3.0		PB H24 H2 6.0 m
PB C36 HN 4.0 m		

[a] The distance constraints are divided into four groups, *intraresidue* (not listed), *sequential, medium-range and long-range backbone*, and *long range,* following the definitions in Section 7.1.

[b] The three-letter amino acid code is at the head of the rows listing NOE's with this residue. In each column the first entry identifies a hydrogen atom in this residue. The second and third entries indicate the one-letter code and the sequence position of a different residue, and the hydrogen atom (or pseudoatom) in this residue to which a distance constraint has been observed. The fourth entry is the distance constraint in Ångstroms, possibly with symbols indicating the correction factors added to allow for the use of pseudoatoms. HN = amide proton, HA = α proton, and so on. M, P, and QR are pseudoatoms, and *m, q,* and *r* are correction terms for the use of pseudoatoms (see Table 10.2). In Asn, the NOE's to the side chain amide protons are referred to the amide nitrogen atom ND, with a correction factor *m*. (From Williamson et al., 1985.)

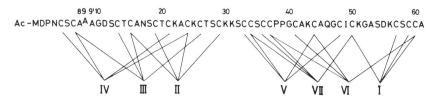

Figure 10.1. Sequence-specific assignments of the COSY connectivities between cysteinyl side chains and the seven metal ions (numbered I–VII in the order of decreasing chemical shift) in $^{113}Cd_7$ metallothionein. (Reprinted with permission from Frey et al., *J. Am. Chem. Soc.*, **107**, 6847; © 1985 American Chemical Society).

3.0 Å for the N–O distance were used, which forces the hydrogen bonds to be approximately linear. Similar constraints were used for two hydrogen bonds with glutamic acid side chains, which had been identified from pH titrations of the amide-proton chemical shifts (Bundi and Wüthrich, 1979b; Ebina and Wüthrich, 1984).

In conjugated proteins, interactions with nuclear or electronic spins located in the nonpeptide components may provide additional NMR constraints on the polypeptide conformation. Nonpeptide components may include metal ions, metal complexes such as heme groups, or rigid organic compounds. They may be intrinsic parts of the protein molecule or may be added for the purpose of the NMR studies, following the much propagated strategy for the use of paramagnetic NMR shift reagents (Dwek, 1973). An illustration is provided by the results obtained with $^{113}Cd_7$ metallothionein-2 using the experiment of Figure 5.40 and similar, additional measurements (Frey et al., 1985b). Combined with the sequence-specific resonance assignments for the polypeptide chain (Wagner et al., 1986), the ^{113}Cd–^1H COSY connectivities identified the Cys residues that are bound to particular ones among the seven metal ions (Fig. 10.1). Assuming tetrahedral coordination of the individual Cd^{2+} ions, with a metal–sulfur bond length of 2.6 Å, the ^{113}Cd–^1H scalar couplings were translated into distance constraints suitable for use in distance geometry computations (Braun et al., 1986). Their distribution along the amino acid sequence (Fig. 10.1) shows that these metal–Cys bonds impose stringent constraints on the entire polypeptide chain. They clearly divide the chain into two domains, as was previously inferred from ^{113}Cd–^{113}Cd spin decoupling experiments, which showed that the seven metal ions are grouped into two clusters containing, respectively, CdII, CdIII, and CdIV and CdI, CdV, CdVI, and CdVII (Armitage and Otvos, 1982).

10.2. MODEL BUILDING WITH INTERACTIVE COMPUTER GRAPHICS

The short, NOE-observable distances constituting the major NMR information on spatial protein structures may connect hydrogen atoms located anywhere in the sequence. Therefore, a graphics program for structure determination from such data must have the entire molecule available at all times. (This contrasts with the process of fitting a polypeptide chain to the electron density map in crystallographic structure determinations, where only a short peptide segment needs to be considered at any given moment.) CONFOR (Billeter et al., 1985) is a program written for this purpose. It can simultaneously display and store one or several molecules. (In the present implementation of the program with a PDP11/34 computer, up to 5 molecules with a total number of 1000 atoms can be handled.) All dihedral angles defined in these molecules are accessible for changes at any time. Violations of upper distance limits obtained by ^{1}H NMR, or of lower limits imposed by the atomic van der Waals volumes are visualized in the protein structure and guide the user in making conformation changes. Complex changes of the conformation can be performed in real time by simultaneous variation of up to eight dihedral angles anywhere in the molecule. In the following, the performance of CONFOR is used to illustrate the model building approach for structure determination from NMR data.

Figure 10.2 describes NMR data obtained for the *lac* headpiece. The sequence locations of three helices were determined using the methods of Chapter 9 (Zuiderweg et al., 1983a). Additional, long-range NOE's were found to connect hydrogen atoms located in different helices (Fig. 10.2). In the treatment of this molecule with CONFOR, the structure determination was reduced to a study of the relative spatial arrangement of the three helices I, II, and III, which were kept in the standard α helix conformation by fixing the dihedral angles at $\phi = -57°$ and $\psi = -47°$. Each of a total of 28 long-range NOE's was attributed an upper distance limit of 4.0 Å, whereby all NOE's with side chain protons were referred to the βC position, with an appropriate correction of the distance constraint. The problem to be solved was to adjust the torsion angles ϕ and ψ in the peptide segments between the

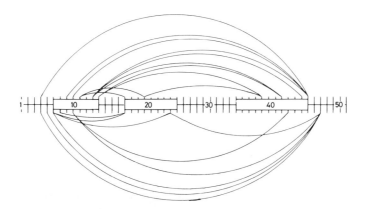

Figure 10.2. Schematic presentation of the amino acid sequence of *lac* headpiece, with three boxes identifying α-helical regions. The curved lines connect residues between which one or several long-range NOE's were observed (from Zuiderweg et al., 1984b).

helices, so that all 28 distance constraints would be satisfied. In a first step, only the segment from residues 6 to 25 was considered, which includes the first two helices and the intervening peptide segment 14–16 with the 6 backbone torsion angles ϕ_{14}–ψ_{16}. Figure 10.3 presents a snapshot from this phase of the project. The dotted arrows indicate the following residual NOE violations in this molecular segment: Leu-6 βC–Tyr-17 βC, Tyr-7 αH–Tyr-17 βC, Ala-10 βC–Tyr17 αH, and Ala-10 βC–Tyr-17 βC. (In reality, a color display with different coloring of helix I, helix II, residues 14–16, and the arrows would be used.) Subsequently the torsion angles ϕ_{14}–ψ_{16} could be adjusted so that all arrows became very short or disappeared. At this point the peptide segment 26–45 (including helix III from residues 34 to 44) was added and the fitting process continued by combined variation of up to 8 torsion angles in the segments 14–16 and 26–33. The helix topology that was found to satisfy all distance constraints is displayed in Figure 10.4.

A similar model building approach could in principle be followed using mechanical molecular models instead of a computer-graphics system (Zuiderweg et al., 1984a,b). Major difficulties tend to arise, however, when trying to keep track of multiple distance violations simultaneously.

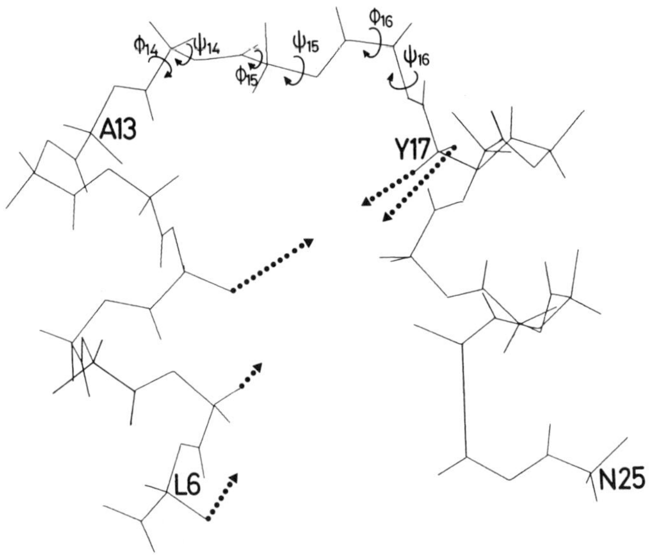

Figure 10.3. View of the picture system screen when the computer-graphics program CONFOR is used for structure determination. The residues 6–25 of the *lac* headpiece (Fig. 10.2) are shown, with the two helices 6–13 and 17–25. The backbone carbonyl oxygens and the side chain atoms beyond βC have been omitted. The dotted arrows indicate NOE violations. They are attached to the atoms involved in the violations and indicate the directions in which the atoms should be moved to meet the NOE constraints. The curved solid arrows identify the torsion angles that could be varied to generate conformation changes (see text).

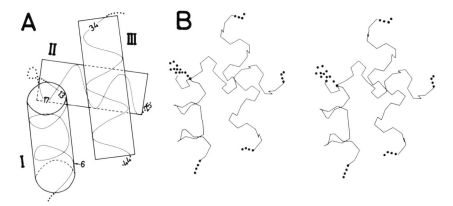

Figure 10.4. Spatial arrangement of the three helices in the *lac* headpiece. *(A)* Mono view, the helices are represented as cylinders. *(B)* Stereoview, all backbone atoms N, αC, and C' are drawn (from Zuiderweg et al., 1984b).

For practical reasons, computer graphics is primarily useful for relatively low-dimensional problems. In the example of Figures 10.2–10.4, low dimensionality was achieved by a drastic simplification of a complex system. As a consequence, no structural details beyond the approximate topology of arrangement of the three helices could be expected from this approach. An alternative use of computer-graphics modeling is for detailed investigation of strictly limited regions, which may be singled out from macromolecular structures. An example is the structural entity consisting of the heme group and the axially bound methionine side chain in cytochromes c (Fig. 6.7*B*). In the two cytochromes shown in Figure 10.5, complete sets of NOE build-up curves were obtained for all combinations of methionine protons among themselves and with the four heme meso protons (an example is shown in Fig. 6.7*A*). This includes NOE's manifesting ^{1}H–^{1}H distances fixed by the covalent structure, as well as unknown distances characterizing the spatial orientation of the methionine side chain relative to the heme plane. Since direct evidence for high rigidity of this structural element was obtained, use of Eq. (6.10) allowed the determination of the ^{1}H–^{1}H distances with much higher accuracy than the distance constraints in Table 10.1 (for most distances to ± 0.1 Å). A structural analysis with CONFOR resulted in stereospecific assignments of the methionine methylene protons and a characterization of the side chain conformation of the axial methionine at atomic resolution (Senn et al., 1984). A particular feature seen in Figure 10.5 is the different chirality at the iron-bound methionine sulfur atom in horse ferrocytochrome c and in *P. aeruginosa* ferrocytochrome c-551.

Structural studies with model building approaches using principles similar to those outlined here were described for various small proteins (e.g., van de Ven et al., 1984; Wemmer and Kallenbach, 1983).

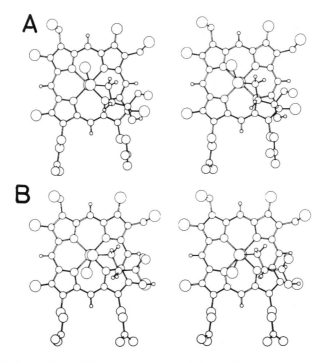

Figure 10.5. Stereoview of the heme group and the axial methionine in the solution conformations of *(A)* horse ferrocytochrome *c, (B) Pseudomonas aeruginosa* ferrocytochrome *c-551* (from Senn et al., 1984).

10.3. DISTANCE GEOMETRY CALCULATIONS

Distance geometry has long attracted interest in different areas of mathematical research (Blumenthal, 1970). Applied to macromolecules, distance geometry algorithms can be used to obtain the Cartesian coordinates of spatial molecular structures that are consistent with a predetermined set of intramolecular distances (Havel et al., 1983). Early studies of proteins used drastically simplified polypeptide structures, with each residue represented by a sphere at the αC position (string of pearls). Using αC–αC distances obtained from the crystal structures, distance geometry calculations with these simple models provided important insights into theoretical aspects of protein folding and the feasibility of reproducing known conformations from suitable sets of intramolecular distances (Crippen, 1977; Kuntz et al., 1976, 1979; Sippl and Scheraga, 1985; Wako and Scheraga, 1981). On a different line, triangulation techniques were used for determination of the spatial arrangement of multimolecular assemblies from distance measurements with neutron interference (Moore et al., 1977; Moore and Weinstein, 1979).

A distance geometry algorithm for structure determinations from NMR data must meet a number of new requirements. These are primarily dictated by the nature of the NOE constraints, representing upper limits on intramolecular distances that are short compared to the overall molecular dimensions of even the smallest proteins. The structure determination then depends critically on complete and faithful representation of the additional constraints imposed by the van der Waals volumes of the individual atoms or groups of atoms and by the limited flexibility of the covalent polypeptide structure. Therefore, the input must contain a complete description of the covalent structure, including the chirality at the asymmetric carbons, and provision must be made to incorporate the pseudoatoms (Table 10.2). This extends the mathematical frame of the embedding procedures used in the classical metric-matrix distance geometry approach (Havel et al., 1983). In practice, the inherent properties of the covalent polypeptide structure correspond to a much larger number of distance and chirality constraints than those typically obtained from the NOE measurements. As a consequence, with increasing molecular size the time required for such calculations and the memory requirements become quite formidable. The first program written for structure determination using NOE distance constraints (Braun et al., 1981) had a size limitation to molecules with approximately 150 atoms, which corresponds to a decapeptide (or a polypeptide chain with 75 residues consisting of two pseudoatoms each). Figure 10.6 shows the conformation of

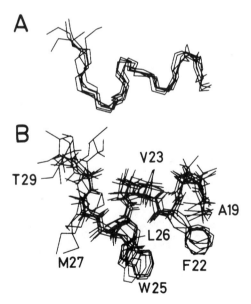

Figure 10.6. Conformation of the residues 19–29 in micelle-bound glucagon. Six conformers determined in separate distance geometry calculations from the same NMR data were superimposed for minimum RMSD. (A) Backbone. (B) Complete structure for the identified residues; for the others the side chain is represented by βC (from Braun et al., 1983).

the C-terminal decapeptide segment in micelle-bound glucagon (Fig. 5.11), which was determined with this program. The size limit could subsequently be increased to polypeptide chains with up to approximately 100 residues, using a new program, DISGEO (Havel and Wüthrich, 1984). DISGEO uses fundamentally the same approach as the earlier calculations for micelle-bound glucagon, with significant improvements including a decomposition of the embedding process into two successive, more tractable calculations, and more efficient storage of the variables. Even with these improvements, a structure determination for a small protein still constitutes a substantial calculation. As an illustration, Figure 10.7 shows the backbone fold in the solution conformations of BUSI determined by DISGEO from two somewhat different input data sets (see Section 10.4).

A different avenue to reduction of the computing time and the memory requirements for finding structures that are compatible with a set of NOE distance constraints is used in the program package DISMAN (Braun and Gō, 1985). This algorithm works in the space of variable torsion angles (Fig.

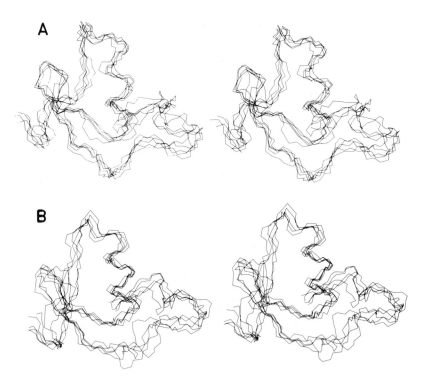

Figure 10.7. Stereoview of the backbone conformation of BUSI in aqueous solution. Five conformers determined in separate distance geometry calculations with DISGEO from the same NMR data were superimposed for minimal RMSD. *(A)* Computed using NOE constraints only. *(B)* Constraints from NOE's, spin–spin couplings $^3J_{HN\alpha}$, and hydrogen bonds (from Williamson et al., 1985).

7.2) instead of a metric space, standard covalent geometry is kept fixed during the calculation, and it uses *variable-target functions* to optimize the molecular conformation.

The following are some general comments on the ways a distance geometry algorithm accepts the input of distance constraints from NOE measurements. They show that distance-geometry is quite well taylored for use with NMR data:

1. As an explicit part of the input for computations in metric space and implicit in the fixed covalent structure in torsion angle space, precise distance and chirality constraints characterize the covalent polypeptide chain. For proteins with 60 amino acids there are approximately 4000 such covalent constraints. An important feature of all constraints obtained from the primary structure is that they are *short range* in nature, since the atoms involved are always covalent neighbors in the molecule. In contrast, NOE distance constraints are often between atoms that are separated by many covalent bonds. Therefore, in spite of their comparatively small number (optimally several hundred NOE's in a protein with 60 amino acid residues) NMR distance constraints play a dominant role in the determination of the global conformation of the polypeptide chain (Havel and Wüthrich, 1985).

2. In structural interpretations with distance geometry the experimental NOE constraints (Tables 10.1 and 10.3) are best interpreted as a range of *equally probable* values for the distances in question (Braun et al., 1981, 1983; Havel and Wüthrich, 1984). This provides for a proper treatment of the NMR data, since the uncertainties in the experimental distance measurements would presently be difficult to account for by a probability distribution. "Pure" distance geometry would then attempt to find an exact fit to these *imprecise but completely correct* data, and ideally a structure obtained by a distance geometry calculation should have no violations of the distance constraints imposed by the experimental data. In practice, finding these structures involves also a numerical optimization versus an error function, and each solution typically includes a certain number of small, residual distance violations (Braun et al., 1981, 1983; Havel and Wüthrich, 1985; Williamson et al., 1985).

3. It lies in the nature of distance geometry calculations using NMR input as described in *2,* that they cannot provide a single, most probable solution. Rather, when presented with relatively wide distance ranges as input (Table 10.1), the goal of distance geometry is to obtain a sampling that is random in the thermodynamic sense of the set of all structures that are consistent with these data. Figures 10.6 and 10.7 present superpositions of groups of conformers obtained as solutions in different computations with the same data input. For all three structures shown, all conformers display the same global features but differ in local aspects of the spatial fold. Each individual conformer in these groups has the same probability of representing the structure of the protein investigated.

4. An important point is to establish criteria for judging the quality of the protein structures determined with distance geometry (see also Section 10.4 and Chapter 14). On the technical side, there are two obvious quantities to be used. The first is that the residual violations of the constraints imposed by the covalent structure and the NMR data should be minimal for an acceptable structure. For a representative conformer of BUSI in the group of Figure 10.7*A*, there was 1 residual violation of 0.6 Å, 3 of 0.5 Å, 6 of 0.3 Å, 4 of 0.2 Å, 48 of 0.1 Å, and approximately 40% of the total of 3700 constraints were violated in the extent of 0.01 to 0.1 Å. Thereby the violations >0.05 Å included very few covalent constraints. (For technical reasons the structures contain in addition certain short contacts between vicinal atoms, and individual bond angles may deviate from the standard geometry for amino acid residues. These energetically unfavorable local structural features were subsequently improved by a standard energy minimization; see Section 10.5.) Second, in a structure determination from a good experimental data set the root-mean-square distances (RMSD) between different conformers computed from the same data should be small. Representative numbers for the RMSD's between the backbone atoms in different pairs of conformers of Figure 10.7*A* are 1.6, 1.8, 2.0, and 2.1 Å. The corresponding RMSD's for the complete structures including amino acid side chains are 2.8, 2.7, 2.8, and 3.1 Å. If this is warranted by the accuracy of the input data, such comparisons can be extended to other properties, for example, certain torsion angles (Havel and Wüthrich, 1985). (Obviously, any conclusions based on the extent of structural differences within the group of conformers representing the result of the structure determination rely on the assumption that the distance geometry algorithm performs a thorough search of the entire conformation space.)

10.4. PROTEIN STRUCTURES DETERMINED WITH NMR AND DISTANCE GEOMETRY

Initial applications of the distance geometry approach for structural interpretation of NMR data include determinations of new polypeptide conformations (Figs. 10.6 and 10.7; Arseniev et al., 1984a,b; Braun et al., 1981, 1983, 1986; Brown et al., 1982; Kline et al., 1986; Williamson et al., 1985) as well as studies on the fundamental problem of the correlations between given input data and the structures computed from them (Havel and Wüthrich, 1984, 1985; Braun and Gō, 1985). In the following, we first address this latter question. Then, using BUSI as an illustration, initial experience in combining distance geometry calculations with interactive techniques for structure determination from experimental NMR data is described. Finally, in Section 10.4.3 the solution conformation of Tendamistat determined with the program DISMAN is presented.

10.4.1. Evaluation of the Combined Use of NMR and Distance Geometry for Protein Structure Determination

A NMR data set for protein structure determination contains a precisely known number of constraints (Table 10.3), and the precision of these constraints is also clearly defined (Tables 10.1 and 10.3). However, there is no straightforward correlation between the number and accuracy of the distance constraints and the computed structure, since the latter also critically depends on the distribution of the constraints along the polypeptide chain (Havel et al., 1979; Wako and Scheraga, 1981). A more reliable basis for evaluating the performance of distance geometry for the structural interpretation of NMR data and for formulating guidelines by which the collection of the NMR input for protein spatial-structure determinations can be efficiently performed, comes from an empirical approach using distance constraints derived from a protein crystal structure (Havel and Wüthrich, 1984, 1985). In this procedure any set of distances used to simulate experimental NMR data is a priori consistent with the protein crystal structure, and comparison of the computed structures with the crystal structure provides a standard by which results obtained with real experimental NMR data can be judged.

For such model computations with DISGEO, the crystal structure of BPTI (Deisenhofer and Steigemann, 1975) was used as the source for the input data. After attachment of hydrogen atoms, this crystal structure contains 508 interresidue ^{1}H–^{1}H distances shorter than 4.0 Å. These were translated into NOE distance constraints of the type listed in Table 10.1 by attributing strong, medium, and weak NOE's, respectively, to distances in the ranges 2.00–2.49 Å, 2.50–2.99 Å, and 3.00–4.00 Å. To prepare data sets simulating realistic NMR situations, only those 356 of these 508 distances were retained, which one could expect to observe in real NOESY experiments. These are all distances corresponding to strong NOE's, and the medium and weak NOE's that involve amide protons or aromatic protons and would therefore be well resolved (Fig. 6.8). In data set A (Table 10.4), weak, medium, and strong NOE's were distinguished throughout, in set B only for the sequential constraints, and in set C the sequential constraints were not included and for all medium and long range constraints the distance limit was 4.0 Å. Set D simulates a situation corresponding to more accurate distance measurements, with all 356 constraints confined to a range of ±0.5 Å about the actual distance in the crystal structure. Set E corresponds to NOESY data recorded with poor S/N, so that only strong and medium NOE's would be observed; otherwise the data handling corresponds to that in set B. With all data sets, pseudoatoms (Table 10.2) were used where applicable. The global shape and the overall backbone fold of the structures computed with each of these sets of input data coincide closely with the crystal structure (Fig. 10.8). The following are some important conclusions from these studies:

TABLE 10.4. Simulated Sets of NOE Distance Constraints Derived from the Crystal Structure of BPTI[a]

Data Set (Total Constraints)	Sequential		Medium and Long Range	
	Number	Limit (Å)	Number	Limit (Å)
A (356)	38	≤ 2.5	35	≤ 2.5
	44	≤ 3.0	53	≤ 3.0
	40	≤ 4.0	146	≤ 4.0
B (356)	38	≤ 2.5	234	≤ 4.0
	44	≤ 3.0		
	40	≤ 4.0		
C (234)			234	≤ 4.0
D (356)	122	± 0.5	234	± 0.5
E (170)	38	≤ 2.5		
	44	≤ 3.0	88	≤ 4.0

[a] The table contains two separate listings for sequential constraints and for medium-range and long-range constraints. The distances are constrained between 2.0 Å and the upper limits indicated in the table, except for data set D, where the distances are constrained within ±0.5 Å about the actual distance in the crystal structure.

1. The dimensions of the computed structures are more precisely defined than what could be achieved with hydrodynamic or thermodynamic studies of the protein solution, even though the longest distance constraints in the input correspond only to a fraction of the molecular diameter.

2. For practical purposes it is of interest that the determination of the global polypeptide fold does not critically depend either on quantitation of the medium- and long-range distance constraints (Fig. 10.8A and B) or on inclusion of the sequential distance constraints (Fig. 10.8C). However, a more precise structure determination would be possible if more accurate distance measurements were available (Fig. 10.8D).

3. Reduction of the total number of medium-range and long-range constraints can result in a markedly less precise global structure determination (Fig. 10.8E).

4. A structure determination from an incomplete set of NOE distance constraints may be substantially improved by inclusion of supplementary constraints corresponding to measurements of $^3J_{HN\alpha}$ and identification of

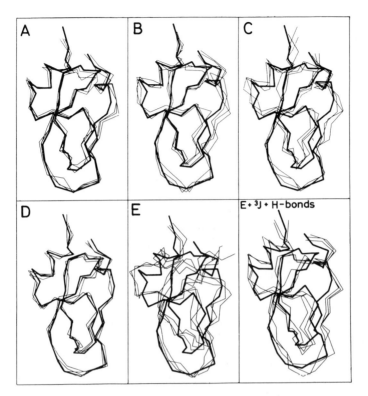

Figure 10.8. BPTI structures computed with DISGEO from distance constraints obtained from the crystal structure. The input data sets A–E are described in Table 10.4. Three computed conformers are superimposed for minimum RMSD on the crystal structure (heavier line), from which the input data were derived (from Havel and Wüthrich, 1985).

hydrogen bonds, as described in Section 10.1 (Fig. 10.8E and $E + {}^3J +$ H-bonds).

   5. The local conformation on the level of the torsion angles ϕ, ψ, and χ^i is only vaguely defined by loose NOE distance constraints of the type described in Table 10.1. Its determination will require either the use of more precise measurements of the short-range distances or direct determination of the dihedral angles about single bonds, or possibly novel concepts of structure refinements within the confines of the long-range NMR constraints.

Overall, the general implications are that in the determination of protein solution structures by NMR and distance geometry, the global conformation is accessible through the imprecise medium- and long-range constraints presently available from NOESY experiments with small proteins, and that work on the global and local conformational features will represent two distinct phases requiring different experimental measurements and different computational procedures.

Similar tests using the crystal structure of BPTI were conducted with the program DISMAN, with overall conclusions generally coinciding closely with those reached in the earlier studies with DISGEO (Braun and Gō, 1985).

10.4.2. Bull Seminal Inhibitor (BUSI). Distance Geometry Using DISGEO Supported by Interactive Decision Making

Experience gained with BUSI (Williamson et al., 1985), BPTI (Fig. 10.8), and other proteins indicates that distance geometry calculations with NOE distance constraints using algorithms such as that implemented in DISGEO might profitably be employed in conjunction with certain interactive manipulations of the data or the resulting molecular structures. For example, one might find in the course of the structure determination that certain low-precision NOE constraints can be substituted by more stringent confines, or inspection of the molecular structure might indicate obvious conformation changes to evade local minima during the optimization. The type of decisions thus called for is further illustrated with the following details from the structure determination of BUSI:

1. The initial input of NOE distance constraints (Table 10.3) was modified by two interactive decisions: A cis peptide bond identified by a sequential connectivity $d_{\alpha\alpha}$ was introduced directly as such, and three disulfide bonds, which were located by NOE's linking the individual cysteines, were fixed directly by imposing a range of 2.0 to 2.1 Å on the S–S distance, and of 3.0 to 3.1 Å on the S–βC distances across each bridge.

2.. In DISGEO, the initial embedding uses a *substructure* of the polypeptide chain and is relatively inexpensive in terms of computation time. Twenty such embeddings were performed and in another interactive step those five embeddings that converged best were selected for the subsequent, much longer calculations. Figure 10.7A shows the five conformers obtained from this process.

3. The 5 conformers in Fig. 10.7A all contain helical and β-sheet secondary structures in identical sequence locations as was previously found by the methods described in Chapter 9 (Fig. 9.3). At this point it was decided to supplement the input of NOE distance constraints by more precise hydrogen bond and $^3J_{HN\alpha}$ constraints, as described in Section 10.1. Five additional conformers computed from this amended data set (Fig. 10.7B) are very similar to those obtained from the NOE constraints alone, except that the helical structures are more precisely defined.

4. The structures in Figure 10.7 contained numerous eclipsed rotamers about single bonds, which is a consequence of some features built into

DISGEO to improve convergence (Havel and Wüthrich, 1984, 1985). These energetically unfavorable local conformations were subsequently improved by a combination of interactive adjustments using CONFOR (Section 10.2) and an energy minimization (see Section 10.5, Fig. 10.10.).

10.4.3. Tendamistat. Distance Geometry Using DISMAN Supported by Interactive Optimization of Intermediate Structures

For the determination of the solution conformation of Tendamistat (Fig. 10.9) the program DISMAN was employed (Kline et al., 1986). Experience gained with this project indicates that tertiary structure determination with restrained structure optimization procedures in torsion angle space, such as DISMAN, can also benefit from support of the automated computational procedure by interactive manipulations. For Tendamistat this involved interactive optimization of the conformers obtained at the early steps of the variable-target function. Again, the program CONFOR was used for inspection of the structures during the interactive parts of this structure determination. It remains to be seen to what extent this laborious examination and optimization of intermediate structures can be automated in future improved algorithms.

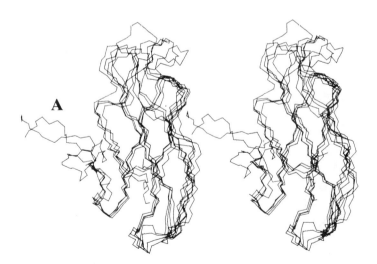

Figure 10.9. Conformation of Tendamistat in aqueous solution. *(A)* Stereo view of four conformers resulting from four DISMAN computations using the same NMR data. Only the bonds connecting the backbone atoms N, αC, and C' are drawn. The four conformers were superimposed for minimum RMSD. *(B)* Polypeptide backbone fold of one of the conformers in *(A)* drawn by CONFOR. The arrowed ribbons represent β sheet strands, and the lightning bolts are disulfide bonds. *(C)* Stereo view showing a superposition of the four conformers of *(A)* in the active site region Ser-17–Trp-18–Arg-19–Tyr-20, including the amino acid side chains (from Kline et al., 1986).

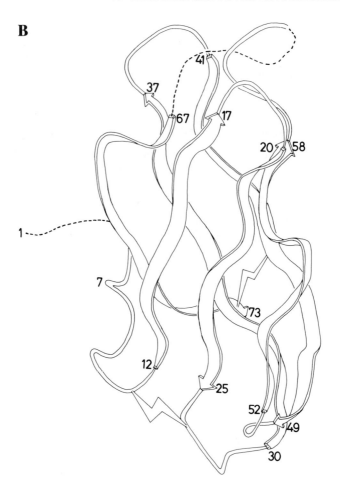

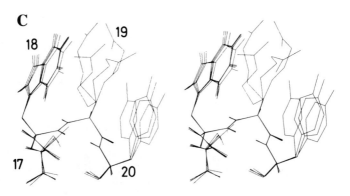

Figure 10.9. (*Continued*)

In the solution conformation of Tendamistat computed with DISMAN (Fig. 10.9A and B) two antiparallel β sheets can readily be recognized, which were previously identified using the pattern recognition approach described in Chapter 9 (Fig. 9.6). These two β sheets form a *greek-key β barrel* with *a* +1, +3, −1, −1, +3 topology (Richardson, 1981). In the active site, which consists of the peptide segment 17–20 in the upper right of the drawing in Figure 10.9B, the side chain of Arg-19 is sandwiched between the aromatic rings of Trp-18 and Tyr-20. In addition to the NOE's between Arg-19 and the two aromatic rings, this structure is also manifested in high field ring current shifts of the side chain ¹H NMR lines of Arg-19.

10.5. STRUCTURE REFINEMENTS

In Section 10.4 it was pointed out that NMR determination of global and local conformational properties of proteins will probably have to be treated in separate phases of a structure determination, which may involve different experiments and different computational procedures. Local conformation on the level of dipeptide segments and individual amino acid side chains could a priori be determined only vaguely by the loose distance constraints used as input for determination of the global polypeptide fold (Tables 10.1 and 10.3). However, in apparent contrast, the spatial arrangement of the amino acid side chains in the interior of the computed structures turned out to be rather well defined. As an illustration, Figure 10.10 shows a superposition of an interior region in one of the BUSI conformers of Figure 10.7B with the corresponding peptide segment in the crystal structure of a homologous protein. The coincidence of most of the side chains is striking, even in some

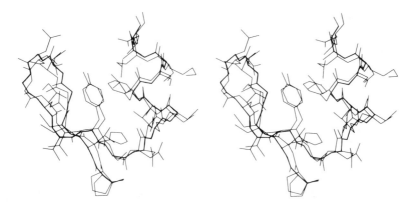

Figure 10.10. Stereoview of a superposition for minimal RMSD of the backbone atoms of the polypeptide segment 23–42 in the solution conformation of BUSI and in the crystal structure of domain 3 of the Japanese quail ovomucoid (from Williamson et al., 1985).

locations containing different residues in the two proteins. Similar observations were made when comparing details of the computed BPTI structures in Figure 10.8 with the crystal structure and with conformations computed from experimental NMR data (unpublished data). The side chains of residues 18–20 in Tendamistat are quite well constrained by the NMR data (Fig. 10.9C), even though they are partially exposed on the protein surface. These observations indicate that packing constraints imposed by the global polypeptide fold are largely responsible for the arrangement of the amino acid side chains, in particular in interior regions of the computed structures.

Observations such as those in Figure 10.10 indicate that protein structures with improved local conformational properties might be obtained not only from accurate measurements of intraresidue and sequential ^{1}H–^{1}H distances and vicinal spin–spin coupling constants, but also from restrained structure refinements within the confines of the experimentally determined global solution conformation. Structure refinements could not be applied before determination of the global polypeptide fold, and therefore it is quite natural that only little actual work has so far been done. This includes the aforementioned improvements of the BUSI structures from DISGEO by removal of eclipsed rotamers (Williamson et al., 1985), and routinely the protein conformations were qualitatively checked against the original NOESY spectra, making certain that there are no experimental NOE's that would correspond to ^{1}H–^{1}H distances longer than 5.0 Å in the computed structures (Braun et al., 1983; Williamson, 1985). A more extensive refinement procedure, using constrained energy minimization and molecular dynamics methods, was applied to the *lac* headpiece, using a molecular model obtained by the methods of Section 10.2 (Fig. 10.4) as the starting structure (Kaptein et al., 1985). A restrained molecular dynamics algorithm was also applied for computation of the conformation of a small polypeptide with 17 amino acid residues from similar NMR data to those in Tables 10.1 and 10.3, using random starting structures (Clore et al., 1985c). The authors point out that the large size of such computations, when compared to distance geometry, will be a limiting factor for similar use with larger molecules. It remains to be seen whether restrained molecular dynamics algorithms will in the future represent an *alternative* to model building (Section 10.2) and distance geometry for determination of the global polypeptide fold, or whether their application is rather indicated for structure refinements in *conjunction* with these techniques.

Obviously, there is much work to be done. This will have to include more accurate analysis of the NOESY data from model computations of the spin-relaxation pathways in the global structures determined by distance geometry techniques (see Section 6.4), which will allow structure refinement against the original NMR data. (This will probably include obtaining stereospecific assignments for a sizable portion of the amino acid side chains.) Efficient methods for *structure regularization* (Braun and Gō, 1985) and for *minimization of the conformational energy* will be needed, and quite possibly further systematic studies on the packing of polypeptide chains under the constraints of semiquantitative NOE data (Tables 10.1 and 10.3) could yield important insights. In addition to providing improved structures for individual proteins, the application of such techniques should allow us to further investigate just how accurately and to what extent (see also Chapter 14) the conformations of noncrystalline proteins can be determined from NMR data. [For example, it will be of considerable interest to investigate if, during refinement, there is a trend toward convergence or divergence among the different conformers obtained from repeated distance geometry calculations with the same input (Figs. 10.6–10.9).]

PART III
RESONANCE ASSIGNMENTS AND STRUCTURE DETERMINATION IN NUCLEIC ACIDS

CHAPTER 11
NOE-Observable ¹H–¹H Distances in Nucleic Acids

This chapter follows a format similar to that used in Chapter 7 for proteins: Standard nucleic acid conformations are described and analyzed for short proton–proton distances to establish a basis for applications of NOE measurements in resonance assignments and spatial structure determination.

11.1.   NOTATION FOR ¹H–¹H DISTANCES IN NUCLEIC ACIDS

The common nucleic acid conformations contain two groups of closely spaced hydrogen atoms (Fig. 11.1). The first group is in the individual DNA strands and comprises the ribose protons and all nonlabile base protons except 2H of adenine. NOE-observable ¹H–¹H distances prevail exclusively between neighboring nucleotides in the sequence. The second group is in the base pairs of duplex structures and comprises the labile base protons and 2H of adenine. All NOE-observable ¹H–¹H distances are between directly adjoining base pairs, but this may include sequential as well as long-range or interstrand distances. For an unambiguous notation (which also distinguishes between sequential distances observed in D_2O and in H_2O) we distinguish four types of distances between protons A and B:

1. Intranucleotide distances are indicated by $d_i(A;B)$, for example,

$$d_i(8;1'), \qquad d_i(1';2') \tag{11.1}$$

2. Sequential distances between nonlabile protons are denoted by $d_s(A;B)$, where the order of A and B is in the direction from the 5' end to the 3' end of the polynucleotide strand (Fig. 11.1). For sequential distances

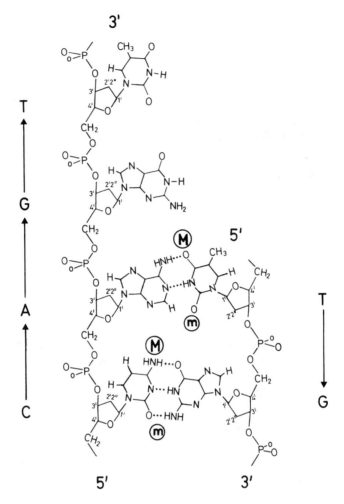

Figure 11.1. Diagram showing an unrolled two base-pair fragment of a DNA double helix and a two base fragment of a single-strand helix. *M* and *m* identify, respectively, the major groove edge and the minor groove edge of the base pairs.

between ribose protons and base protons we write, for example,

$$d_s(1';8), \qquad d_s(1';6), \qquad d_s(1';5), \qquad d_s(1';M) \qquad (11.2)$$

and for sequential base proton–base proton distances

$$d_s(8;8), \qquad d_s(8;6), \qquad d_s(8;5), \qquad d_s(8;M) \qquad (11.3)$$

The letter M is used for the methyl protons of T.

3. Interstrand distances between protons within a base pair are denoted as $d_{pi}(A;B)$. A and B are defined by the nucleotide type and the atom position, whereby A6 and C4 stand for the amino proton that is hydrogen-bonded in Watson–Crick base pairs (Fig. 3.9).

$$d_{pi}(T3;A2), \qquad d_{pi}(T3;A6), \qquad d_{pi}(G1;C4) \qquad (11.4)$$

It is often of interest to further identify the sequence locations of A and B. Examples for tRNA and for a DNA duplex are

$$d_{pi}(U8,3;A24,2), \qquad d_{pi}(A3+,2;T13-,3) \qquad (11.5)$$

where the two strands in the DNA double helix are arbitrarily labeled $+$, and $-$, respectively.

4. *Distances between labile base protons or A2H in adjoining stacked base pairs* are denoted $d_{ps}(A;B)$. The protons A and B are identified by atom position, nucleotide type, sequence position, and, if applicable, specification of the polynucleotide strand. $d_{ps}(A;B)$ includes sequential as well as long range or interstrand distances. For example, for tRNA,

$$d_{ps}(U8,3;G9,1), \qquad d_{ps}(U8,3;C23,4) \qquad (11.6)$$

or for a DNA duplex,

$$d_{ps}(T1+,3;G2+,1), \qquad d_{ps}(A3+,2;G12-,1) \qquad (11.7)$$

In describing the connectivities used for sequential resonance assignments of the nonlabile protons (Chapter 13), a compact notation indicates that the distance is either to 6H of pyrimidine or 8H of purine, or to 5H of C (or U) or M of T

$$d_i(6,8;1'), \qquad d_s(1';6,8), \qquad d_s(6,8;5,M) \qquad (11.8)$$

As a variant which is not used in this book, the sequence positions might be included as subscripts to make the distance notations easier to read. The distances defined in Eqs. (11.6) and (11.7) would then read, for example, $d_{ps}(U_8,3;G_9,1)$ and $d_{ps}(A_{3+},2;G_{12-},1)$.

11.2. INTRANUCLEOTIDE ^{1}H–^{1}H DISTANCES

Similar to polypeptides, different polynucleotide conformations can be characterized by *torsion angles* about single bonds (Fig. 11.2). Along the backbone there are six torsion angles per mononucleotide, α–ζ. The ribose ring conformation is determined by five *endocyclic* torsion angles ν_0–ν_4, where ν_3 is about the same bond as δ. The orientation of the base relative to the ribose ring is determined by χ. Intranucleotide ^{1}H–^{1}H distances are dependent on γ, ν_0–ν_4, and χ. These angles are defined as follows: γ involves the atoms 5'O–5'C–4'C–3'C and is equal to the angle formed by the bonds 5'O–5'C and 4'C–3'C. When viewed along the 5'C–4'C bond, $\gamma = 0$ if the bonds 5'O–5'C and 4'C–3'C are eclipsed, and γ is positive if the far bond is rotated

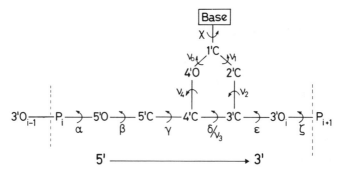

Figure 11.2. Notation for torsion angles about single bonds in the mononucleotide *i* of a polynucleotide chain.

clockwise with respect to the near bond. Corresponding definitions for the origin and the sign hold for the other angles: v_0 (4'C–4'O–1'C–2'C); v_1 (4'O–1'C–2'C–3'C); v_2 (1'C–2'C–3'C–4'C); v_3 (2'C–3'C–4'C–4'O); v_4 (3'C–4'C–4'O–1'C); χ (4'O–1'C–1N–2C for pyrimidines, 4'O–1'C–9N–4C for purines).

For a computation of the angular dependence of the ¹H–¹H distances, the dimensionality of the system can be reduced through substitution of v_0–v_4 by the *pseudorotation phase angle P* (Altona and Sundaralingam, 1972; de Leeuw et al., 1980). Figure 11.3 presents the pseudorotation cycle with a description of different furanose ring puckers. Distances between different furanose ring protons can then be computed as a function of P (Fig. 11.4), those between 1'H and the base protons as a function of χ (Fig. 11.5), and those between the other ribose ring protons and the base protons are dependent on P and χ (Fig. 11.6A–D). Similarly, the distances between the ribose ring protons and 5'H or 5"H can be computed as a function of γ and P (Fig. 11.7). For certain aspects of the interpretation of ¹H NMR measurements it is of interest to know about all short ¹H–¹H distances. Therefore, Table 11.1 additionally lists the short ¹H–¹H distances in the covalent structures of the mononucleotides.

An initial inspection of Figures 11.4–11.7 and Table 11.1 shows that there are a large number of intranucleotide ¹H–¹H distances shorter than 4.5 Å, which should be observable by NOE's. This includes all the distances between protons in the furanose ring, independent of P (Fig. 11.4), and those bewteen $5'CH_2$ and the furanose ring protons for most combinations of P and γ (Fig. 11.7). Most interesting, however, is the fact that there are numerous short contacts between the ribose ring and the base, since these hydrogen-atoms cannot otherwise be connected by scalar coupling in the ¹H NMR spectra (Table 2.5) [except possibly in very short oligonucleotides (Gundhi et al., 1985)]. The NOE connectivities with the ribose ring will be essentially limited to 8H of purines and 6H of pyrimidines, with only a small chance that

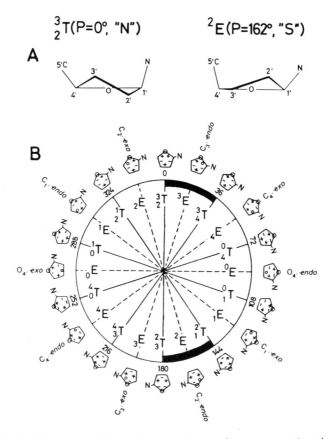

Figure 11.3. (A) Illustration of two furanose ring puckers representing the preferred *north* and *south* conformers. The ring is viewed perpendicular to the plane defined by 1'C–4'O–4'C. Atoms displaced from this plane on the same side as 5'C are called *endo*, those on the opposite side *exo*. The symmetrical twist form $^3_2$T is therefore 2'C-*exo*-3'C-*endo*, and the envelope form ^{2}E is 2'C-*endo*. (B) Pseudorotation cycle of the furanose ring in nucleosides. Values of the pseudorotation phase angle are given in multiples of 36°. Envelope (E) and twist (T) forms alternate every 18°. On the periphery of the cycle, riboses with signs of the endocyclic torsion angles are indicated: +, positive; −, negative; O, 0°. Thick lines inside the circle indicate the preferred pseudorotational regions. (Reprinted with permission from Altona and Sundaralingam, *J. Am. Chem. Soc.*, **94**, 8205–8212; © 1972 American Chemical Society.)

NOE's with 2H of purines or 5H of pyrimidines might also be seen (Figs. 11.5 and 11.6). Outstandingly short distances in B-DNA are $d_i(8;2')$ or $d_i(6;2')$, in A-DNA $d_i(8;3')$ or $d_i(6;3')$, and in Z-DNA $d_i(8;1')$ for G. From Table 11.1 we conclude further that within the nucleotides A, G, and T the labile base protons and A2H are isolated from the other hydrogen atoms not only through the absence of scalar couplings (Table 2.5), but also by long through-space distances.

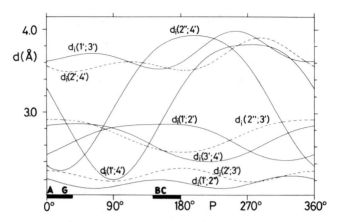

Figure 11.4. Intranucleotide distances between different hydrogen atoms in the deoxyribose ring versus the pseudorotation phase angle P. The heavy lines below the P axis indicate the regions of the preferred ring puckers 3'C-*endo* near 20° and 2'C-*endo* near 160°. A, B, C, and G identify the P-values for A-DNA, B-DNA, and the nucleotides C and G in the Z form of d(CGCGCG)$_2$. This computation and those in the following Figures 11.5–11.7 used the atom coordinates in a standard B-DNA double helix (Arnott and Chandrasekaran, private communication, 1984).

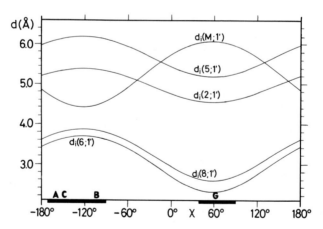

Figure 11.5. Intranucleotide distances between base protons and 1'H versus the torsion angle χ. Preferred χ regions from −90 to −170° (anti) and 40 to 90° (syn) are indicated with heavy lines at the bottom. A, B, C, and G identify the χ-values for A-DNA, B-DNA, and the nucleotides C and G in the Z form of d(CGCGCG)$_2$.

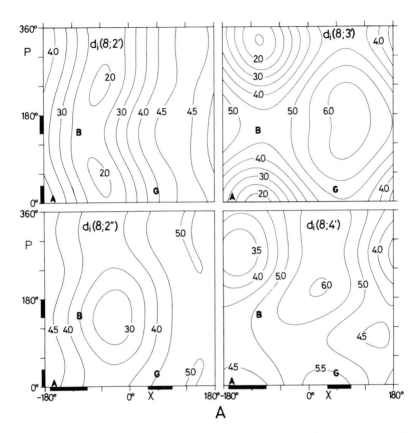

Figure 11.6. P–χ plane with contour lines indicating the distances to the deoxyribose ring protons 2′, 2″, 3′, and 4′ for *(A)* purine 8H, *(B)* purine 2H, *(C)* pyrimidine 6H, and *(D)* cytosine 5H and thymine 5CH₃. The distances from 2″H and 4′H to 5H or 5CH₃ of the pyrimidines are not shown, since they are always longer than 5.0 Å. The preferred regions 3′C-*endo* near 20° and 2′C-*endo* near 160° for P, and syn near 60° and anti near −130° for χ are indicated with heavy lines on the left and at the bottom. A, B, C, and G identify the (P,χ) combinations for A-DNA, B-DNA, and the nucleotides C and G in the Z form of d(CGCGCG)₂.

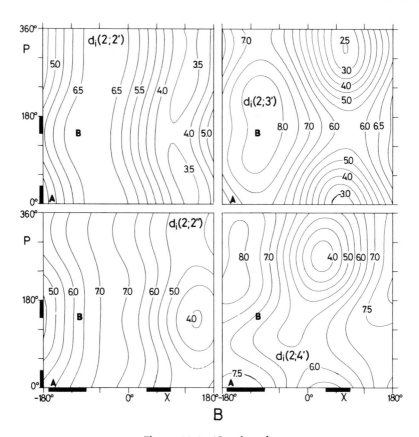

Figure 11.6. (Continued)

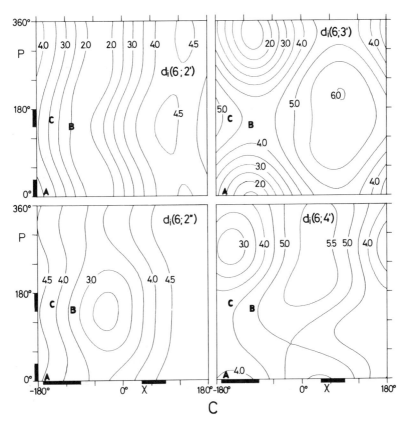

Figure 11.6. (Continued)

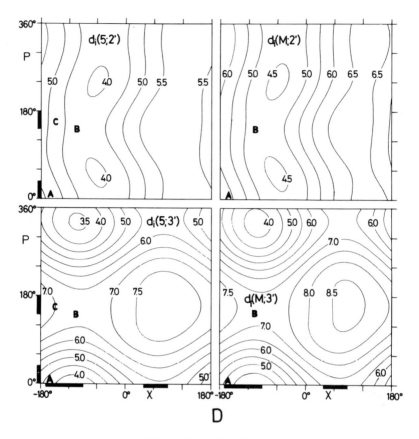

Figure 11.6. (Continued)

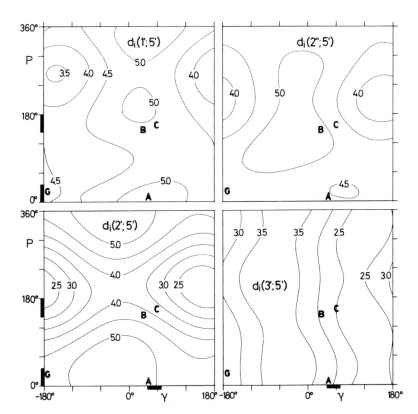

Figure 11.7. P–γ plane with contour lines indicating the distances from 5'H to the deoxyribose ring protons 1', 2', 2", and 3'. (The corresponding distances for 5"H are given by identical contours shifted 120° along the γ axis.) The preferred regions 3'C-*endo* near 20° and 2'C-*endo* near 160° for P and + *synclinal* near 60° for γ are indicated with heavy lines on the left and at the bottom. A, B, C, and G identify the (P,γ) combinations for A-DNA, B-DNA, and the nucleotides C and G in the Z form of d(CGCGCG)₂.

TABLE 11.1. Short Covalent ^{1}H–^{1}H
Distances in the Common DNA
Nucleotides[a]

Distance	Value (Å)
$d_i(2';2'')$	1.8
$d_i(5';5'')$	1.8
$d_i(A2;6NH_2)$	4.4,5.2
$d_i(A6NH;6NH)$	1.7
$d_i(A6NH_2;8)$	4.8,<u>6.1</u>
$d_i(G1;2NH_2)$	<u>2.3</u>,3.4
$d_i(C4NH_2;5)$	2.4,<u>3.6</u>
$d_i(C4NH_2;6)$	4.6,<u>5.3</u>
$d_i(C5;6)$	2.4
$d_i(T3;M)$	4.9
$d_i(TM;6)$	2.9

[a] Computed from the atom coordinates in a standard B-DNA double helix (Arnott and Chandrasekaran, private communication, 1984). Identical distances occurring in several nucleotides are listed only once. For NH_2, the distance to the hydrogen-bonded proton in the Watson–Crick base pairs (Fig. 3.9) is underlined. M denotes 5CH$_3$ of T, and the distances are to the center of the three methyl protons.

11.3. SEQUENTIAL DISTANCES BETWEEN NONLABILE PROTONS IN A-DNA AND B-DNA

Instead of a complete description of the conformation dependence of sequential ^{1}H–^{1}H distances, which would be impractical because of the large number of intervening torsion angles (Fig. 11.2), Table 11.2 lists short sequential distances for three standard nucleic acid conformations. The table presents only a general guide, since the average structures used do not include the sequence-dependent variations observed in experimentally determined DNA conformations (Calladine, 1982; Dickerson, 1983).

TABLE 11.2. Short Sequential Distances between Nonlabile Protons in Three Standard DNA Conformations

Distance[a]	A-DNA[b]	B-DNA[b]	Z-DNA,CG[b]	Z-DNA,GC[b]
d_s (1';5")	4.6 Å	3.0 Å	3.7 Å	–
2';1'	4.0	–	–	–
2';3'	4.1	–	–	–
2';5"	2.7	–	5.0	–
2";2'	–	3.7	–	–
2";5"	2.4	4.0	4.6	–
3';2'	–	–	–	3.3 Å
3';3'	–	–	–	4.5
4';5"	4.1	3.4	2.0	–
1';5	–	4.3		4.8
1';6	4.6	3.5		–
1';8	4.6	3.6	–	
2';5	3.5	3.6		2.8
2';M	3.8	3.4		
2';6	2.0	4.0		3.0
2';8	2.1	3.8	–	
2";5	–	2.7		4.2
2";M	–	2.9		
2";6	3.7	2.2		4.2
2";8	3.9	2.1	–	
3';5	3.3	–		4.2
3';M	3.0	–		
3';6	3.1	4.6		3.4
3';8	3.0	4.6	–	
2 ;1'	4.1	–		
5;5	3.7	4.7		
5;M	3.3	4.7		
M;5	4.3	–		
M;M	3.7	5.0		
6;5	3.9	3.9		
6;M	3.5	3.8		
6;8	4.5	4.8	–	
2;2	4.8	3.6		
8;5	3.8	3.9		4.9
8;M	3.4	3.8		
8;8	4.6	5.0		

[a] Distances are listed if they are shorter than 4.5 Å in at least one conformation, except that those involving 4'H, 5'H, or 5"H are only included if they are shorter than 3.0 Å at least once. Then, values up to 5.0 Å are listed, and a dash indicates values longer than 5.0 Å. For the notation used see Eqs. (11.2) and (11.3).

[b] Regular A-DNA and B-DNA were constructed with standard atom coordinates (Arnott and Chandrasekaran, private communication, 1984). The data for Z-DNA are from the atom coordinates of the single-crystal structure of d(CGCGCG)₂ (Wang et al., 1981).

Table 11.2 contains three parts, namely, the distances between different ribose protons, between ribose protons and base protons, and between different base protons. Quite generally, the NOESY cross peaks linking different ribose protons will be located in spectral regions that are crowded with intranucleotide connectivities between the same proton types (Figs. 11.4 and 11.7). Furthermore, in Table 11.2 all ribose–ribose distances shorter than 3.0 Å involve 5″H, which is particularly difficult to resolve in the NMR spectra. While knowledge of the short sequential distances between ribose protons is indispensable for detailed analyses of spin-relaxation pathways, their use for structure determination by NOE's will therefore be rather limited.

Some general characteristics are readily apparent for the other two groups of distances. Thus, with the single exception of $d_s(2;1')$ for adenine in A-DNA, the short sequential ribose proton–base proton distances are in the direction 5′ to 3′. In many instances these sequential distances are shorter than the intranucleotide distances between corresponding proton types (Fig. 11.6), indicating that they will be prominently manifested in NOESY. Extremely short distances are $d_s(2';6)$ or $d_s(2';8)$ in A-DNA, and $d_s(2'';6)$ or $d_s(2'';8)$ in B-DNA. In both A-DNA and B-DNA, base proton–base proton distances shorter than 4.5 Å prevail between neighboring pyrimidines and in purine–pyrimidine dinucleotides. In addition, in A-DNA the pyrimidine–purine connectivity $d_s(6;8)$ and the purine–purine connectivity $d_s(8;8)$ might be within the range for NOE observation, and in B-DNA there is a unique short contact $d_s(2;2)$ between neighboring adenines.

11.4 SHORT SEQUENTIAL AND INTERSTRAND DISTANCES WITH LABILE BASE PROTONS AND A2H IN A- AND B-DNA DUPLEXES

Tables 11.3 and 11.4 list, respectively, short sequential and interstrand distances with labile base protons and A2H measured in regular A-DNA and B-DNA duplexes. Similar to Table 11.2, these data can only serve as a general guide, since they do not allow for sequence-dependent variations of nucleic acid conformations. The labile protons are primarily of interest when they are involved in hydrogen bonds. Therefore, for each NH$_2$ group only the distance to the proton that is hydrogen-bonded in Watson–Crick base pairs (Fig. 3.9) is included.

The van der Waals interactions between coplanar stacked bases impose a lower limit of approximately 3.4 Å for ^{1}H–^{1}H distances between directly adjoining base pairs. This is clearly manifested in Tables 11.3 and 11.4, where nearly all distances d_{ps} in the NOE-observable range are between 3.4 and 4.3 Å.

In the ^{1}H NMR spectra the resonances of the hydrogen-bonded imino protons are well resolved at the low field end (Figs. 3.8 and 3.10), and therefore short distances between these protons are of particular practical

TABLE 11.3 Short Sequential ¹H–¹H Distances with Labile Base Protons in Three Standard DNA Conformations

Distance[a]	A–DNA	B–DNA	Z–DNA
d_{ps} (A1+,2;A2+,2)	4.8 Å	3.6 Å	
A1+,6;A2+,6	3.0	3.7	
A1+,2;T2+,3	4.3	4.0	
A1+,6;T2+,3	3.9	4.5	
A1+,2;G2+,1	–	4.3	
A1+,2;G2+,2	–	4.2	
A1+,6;G2+,1	4.0	3.7	
A1+,6;C2+,4	3.0	4.2	
T1+,3;T2+,3	3.8	3.5	
T1+,3;A2+,2	–	4.4	
T1+,3;A2+,6	4.5	4.0	
T1+,3;G2+,1	4.8	3.8	
T1+,3;C2+,4	3.8	3.9	
G1+,1;G2+,1	3.8	3.4	
G1+,2;G2+,2	4.8	3.7	
G1+,1;A2+,2	4.3	3.7	
G1+,1;A2+,6	4.2	4.5	
G1+,2;A2+,2	4.2	3.5	
G1+,1;T2+,3	3.4	3.9	
G1+,2;T2+,3	4.3	4.5	
G1+,1;C2+,4	4.0	4.7	4.0
C1+,4;C2+,4	3.0	3.7	
C1+,4;A2+,6	3.3	3.4	
C1+,4;T2+,3	4.1	4.3	
C1+,4;G2+,1	4.6	4.0	–
C1+,6;G2+,1	–	–	3.8
X1+,1';G2+,1	–	–	3.5
X1+,1';G2+,2	–	–	3.5

[a] Distances are included if they are shorter than 4.5 Å in at least one conformation. Then, values up to 5.0 Å are listed, and a dash indicates values longer than 5.0 Å. The notation of Eq. (11.7) was used, with the first nucleotide arbitrarily placed in sequence position 1. For NH_2 groups the distance to the hydrogen-bonded proton is given. In the last two entries, X stands for any of the four nucleotides.

Regular A-DNA and B-DNA was constructed with standard coordinates (Arnott and Chandrasekaran, private communication, 1984) for the duplex with the sequence d(TAATTGGCCAGACTCGT), which contains all possible dinucleotides. The data for Z-DNA are from the atom coordinates of the single-crystal structure of d(CGCGCG)$_2$ (Wang et al., 1981).

TABLE 11.4. Short Interstrand ^{1}H–^{1}H Distances in Three Standard DNA Conformations

Distance[a]	A–DNA	B–DNA	Z–DNA
d_{pi} (A2;T3)	2.8 Å	2.8 Å	
A6;T3	2.5	2.5	
G1;C4	2.5	2.5	2.5 Å
d_{ps} (A2+,2;A1-,2)[b]	4.2 Å	2.9 Å	
A2+,2;A3-,2	–	4.4	
A2+,6;A1-,6	3.2	4.0	
A2+,6;A3-,6	2.7	3.3	
A2+,2;T1-,3	4.0	3.8	
A2+,6;T1-,3	4.1	4.8	
A2+,6;T3-,3	4.1	3.5	
A2+,2;G1-,1	4.7	3.8	
A2+,2;G1-,2	4.9	3.4	
A2+,6;G1-,1	4.3	4.2	
A2+,2;G3-,1	4.7	4.1	
A2+,2;G3-,2	4.7	4.1	
A2+,6;G3-,1	3.9	4.1	
A2+,6;C1-,4	3.2	4.5	
A2+,6;C3-,4	3.0	2.9	
T2+,3;T1-,3	3.6	4.3	
T2+,3;T3-,3	–	4.1	
T2+,3;G1-,1	3.5	3.5	
T2+,3;G1-,2	4.3	3.5	
T2+,3;G3-,1	4.3	3.6	
T2+,3;C1-,4	4.2	–	
T2+,3;C3-,4	4.9	3.9	
G2+,1;G1-,1	4.3	3.4	–
G2+,1;G1-,2	–	4.1	–
G2+,2;G1-,1	–	4.1	–
G2+,2;G1-,2	–	4.1	–
G2+,1;G3-,1	3.6	3.6	3.6
G2+,1;G3-,2	4.5	4.3	4.3
G2+,2;G3-,1	4.5	4.3	4.3
G2+,2;G3-,2	4.3	4.0	4.3
G2+,1;C1-,4	3.9	4.2	
G2+,1;C3-,4	4.4	4.2	
C2+,4;C1-,4	3.5	–	4.2
C2+,4;C3-,4	3.6	2.9	4.9
C2+,5;C3-,5	–	–	4.5
A2+,2;X1-,1'	4.1	–	

[a] Distances are included if they are shorter than 4.5 Å in at least one conformation. Then, values up to 5.0 Å are listed, and a dash indicates

interest. In A-DNA and B-DNA, short imino proton–imino proton distances prevail between all combinations of base pairs.

Overall, similar features for A-DNA and B-DNA emerge from Figures 11.5 and 11.6 and Tables 11.1–11.4. For both, each individual DNA strand contains a continuous train of short intraresidue and sequential ^{1}H–^{1}H distances linking the ribose moieties with the nonlabile base protons, which extends from the 5′end to the 3′end of the strands. In addition, a continuous network of short distances connects the labile base protons and A2H over the whole length of the duplexes. There are thus three parallel linear networks of potential NOE connectivities in the direction of the helix axis. The individual networks are largely isolated from each other by long intervening ^{1}H–^{1}H distances, the only exceptions being the short distances between the protons in positions 4, 5, and 6 of C, and $d_s(2;1')$ and $d_{ps}(A2+,2;X1-,1')$ for adenine in A-DNA. Short interstrand ^{1}H–^{1}H distances involve only the labile base protons and A2H, the only exception being that A-DNA contains a short interstrand contact between A2H and 1′H (Table 11.4).

11.5. SHORT ^{1}H–^{1}H DISTANCES IN THE Z FORM OF d(CGCGCG)$_2$

Tables 11.2–11.4 contain a limited number of entries for Z-DNA, which were computed from the single crystal structure of d(CGCGCG)$_2$ (Wang et al., 1981). These data present an adequate guideline for NOE studies of Z-DNA's containing alternating pyrimidine–purine sequences with the pyrimidines in the anti conformation and the purines in the syn conformation, even if these also contain A and T (Brennan and Sundaralingam, 1985; Wang et al., 1984). They would probably have to be expanded for proper treatment of Z-DNA's with sequences deviating from regular pyrimidine–purine alternation, or with pyrimidines in the syn form and purines in the anti form (Wang et al., 1985).

Figures 11.5, 11.6A, and 11.6C show that similar to A- and B-DNA, there are short intranucleotide ribose proton–base proton distances in Z-DNA, which allow connectivities to be established by NOE's. An outstandingly short distance is $d_i(8;1')$ for G.

d(CGCGCG)$_2$ contains short sequential distances between ribose ring protons and nonlabile base protons only for the segment GC (Table 11.2), and

values longer than 5.0 Å. The notation of Eq. (11.7) is used, where the first base was arbitrarily placed in the position 2+ and the bases 1− and 3− are paired with the bases 1+ and 3+, respectively. In the last distance listed, X stands for any of the four bases. Note that the order of the base types can be inverted for all distances defined in this table. For NH$_2$ groups the distance to the hydrogen-bonded proton is given. Regular A-DNA and B-DNA was constructed as in Table 11.3. The data for Z-DNA are from the atom coordinates of the single-crystal structure of d(CGCGCG)$_2$ (Wang et al., 1981).

short contacts between different labile protons are also limited to this same dinucleotide segment (Table 11.4). At the CG junctions, there are three short sequential distances linking labile with nonlabile protons (Table 11.3), and a short interstrand contact between the 5 protons of C (Table 11.4). Thus, in Z-DNA only the combination of labile and nonlabile protons affords a continuous train of short ¹H–¹H distances in the direction of the helix axis.

11.6. CHARACTERIZATION OF DNA CONFORMATIONS BY HELIX PARAMETERS

The torsion angles defined in Figure 11.2 do not provide an intuitively clear description of the conformation of individual polynucleotide strands, and they do not characterize the formation of duplex structures. These are, therefore, more commonly specified by *helix parameters* (Table 11.5), which characterize features of the individual strands as well as their combination to

TABLE 11.5. Mean Double-Helix Parameters for A-DNA, B-DNA, and Z-DNA.[a]

Helix parameter	A-DNA	B-DNA	Z-DNA	
Handedness	right	right	left	
Bases per turn	10.9	10.0	12.0	
Height per base	2.9 Å	3.4 Å	GC	3.5 Å
			CG	4.1 Å
Pitch	31.6 Å	34.0 Å	45.6 Å	
Glycosyl angle	anti	anti	C	anti
			G	syn
Sugar pucker	3'C endo	2'C endo	C	2'C endo
			G	3'C endo
Base pairing	Watson-Crick	Watson-Crick	Watson-Crick	
Repeating helix unit	1 base pair	1 base pair	2 base pairs	
Twist per base pair	33°	36°	GC	−51°
			CG	− 9°
Axis displacement[b]	4 Å	0	−3 Å	
Major groove	very deep	deep, wide	shallow	
Minor groove	shallow	deep, narrow	very deep	
Base inclination	13°	−2°	9°	
Base roll	6°	−1°	3°	
Propeller twist	15°	12°	4°	

[a] Data from single crystal X-ray studies, adapted from Dickerson (1983), Dickerson et al. (1982), and Saenger (1984). The data for Z-DNA are from d(CGCGCG)₂ (Wang et al., 1981).
[b] Positive numbers indicate a displacement of the helix axis toward the major groove edge of the base pair.

double helices. As a visual guide to the following discussion of these helix parameters, Figure 11.8 presents stereoviews of a B-DNA duplex.

The *handedness* describes the sense of the helix. The *pitch P* relates the number of bases per turn *n* and the height *h* per base along the helix axis: $P = n \cdot h$. The *glycosyl angle* χ is in either one of the two preferred states defined in Figure 11.5. The *sugar pucker* is either north, near 3'C-*endo*, or south, near 2'C-*endo* (Fig. 11.3). All three helices contain Watson–Crick base pairs. Z-DNA is unique in that it contains a *dinucleotide repeat* unit (see also Table 11.2). The *twist* per base pair is equal to $360°/n$, with largely different values for G≡C or C≡G in Z-DNA formed by d(CGCGCG)$_2$. The base pairs are not centered on the helix axis, but are displaced from it to variable extent, depending on the conformation type. Because of this *axis displacement* and the fact that the glycosyl bonds branch off from one side of the base pairs (Fig. 11.1), the outer envelope of the double helices is not cylindri-

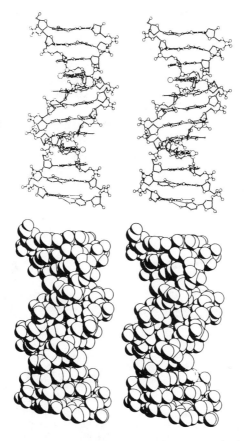

Figure 11.8. Skeletal and space-filling stereodrawings of the B-DNA duplex d(CGCGAATT^{Br}CGCG)$_2$. Atoms in order of descending radius are Br, P, O, N, C (from Dickerson et al., 1982. © 1982 by the AAAS).

cally smooth but displays two grooves, which have variable width and depth in the different forms of DNA. The edge of the base pairs along which the angle between the two glycosidic attachments is less than 180° is called the *minor groove edge,* and the opposite edge of the same base pair is the *major groove edge* (Fig. 11.1). The ribose–phosphate backbones (Fig. 11.1) build the two walls of a major and a minor groove, which wind around the helix with these base-pair edges as the floor of the grooves (Fig. 11.8): For NMR studies it is of interest that different hydrogen atoms are exposed to the solvent in the major and minor groove (Fig. 11.1). For the base pair A=T the major groove edge contains the protons A8H, A6NH$_2$, T5CH$_3$ and T6H, whereas A2H and T3NH are in the minor groove. For G≡C, G8H, C4NH$_2$, C5H, and C6H are in the major groove, and G6NH$_2$ is in the minor groove. The *base inclination* is the angle formed by a straight line through the atoms 6C and 8C of the base pair (Fig. 3.9) and its projection onto a plane perpendicular to the helix axis. The *base roll* is a rotation of the mean plane of the base pair about the axis through the atoms 6C and 8C. These two parameters thus specify deviations of the mean base pair plane from perpendicularity relative to the helix axis. Finally, the two bases in a base pair are not exactly coplanar, as described by the *propeller twist* (Fig. 11.9): If one views a base pair in the direction of the long axis through the atoms 6C and 8C (Fig. 3.9), a clockwise rotation of the nearer base relative to the plane of the other base is considered a positive propeller twist.

Clearly, the contacts between adjacent base pairs and hence the ¹H-¹H distances can be affected when neighboring stacked bases deviate differently from perpendicularity relative to the helix axis. This is illustrated in Figure 11.9, where propeller twist affects the interstrand contacts between purines. For example, the distance d_{ps}(A1+,2;A1−,2) in Figure 11.9A (in the two

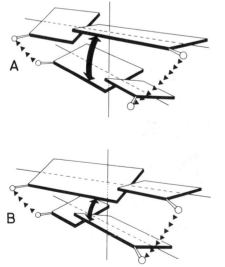

Figure 11.9. Effect of propeller twist on ¹H–¹H distances between stacked purines in the different strands of a double helix. The bases are presented as planks, the carbon atoms 1′ as small spheres, and the backbones as a line of arrows pointing in the 5′-to-3′ direction. The thick arrows indicate the distance between the protons 2H of adenine in adjoining A=T base pairs (from R. E. Dickerson, *Scientific American*, December 1983; copyright I. Geis).

strands, + and −, the nucleotides are numbered in the 5'–3' direction) would be considerably longer than the distance $d_{ps}(A2+,2;A2-,2)$ in the dinucleotide duplex of Figure 11.9B. More generally, in purine–pyrimidine segments positive propeller twist results in close major groove contacts between stacked purines on the different strands of the double helix, and in pyrimidine-purine sequences positive propeller twist causes close contacts between the minor groove edges.

So far the discussions in this chapter made reference almost exclusively to DNA conformations. In as far as RNA's go, present knowledge indicates quite convincingly that the stem regions in tRNA and double-helical structures formed by self-complementary segments in mRNA or rRNA correspond to variants of the A double helix (Dickerson, 1983; Saenger, 1984).

11.7. IMPLICATIONS FOR THE ANALYSIS OF NOESY SPECTRA

The NOESY spectra of nucleic acids contain cross peaks corresponding to intranucleotide, sequential, and long-range or interstrand short distances between the same proton types. For ribose-base connectivities, which are of prime interest, the intranucleotide NOESY cross peaks cannot be identified as such by comparison with COSY. Distinction between intranucleotide and internucleotide ribose proton–base proton NOE's will therefore have to be part of a complete analysis of the NOESY spectra. Similarly, separate identification of sequential, long-range, and interstrand connectivities with labile base protons will also be a major result of the complete spectral analysis. Overall, a quite uniform appearance of the NOESY spectra for different duplex structures is ensured by the fact that most NOE's are between the same proton types. For example, the spectrum of d(CGCGAATTCGCG)$_2$ in Figure 6.9 is representative for B-DNA duplexes.

In Section 11.4 we concluded that A-DNA and B-DNA contain three linear networks of short ^{1}H–^{1}H distances parallel to the helix axis, which are separated from each other by long intervening distances. Therefore, the analysis of NOESY spectra in D$_2$O can rely on the fact that cross relaxation will be confined to protons within the individual DNA strands. Furthermore, once the intranucleotide and sequential NOE's are assigned, there should be no unassigned cross peaks left in the NOESY spectra of DNA duplexes in D$_2$O (with the possible exception of some A2H–A2H or A2H–1'H interstrand NOE's).

For the hydrogen-bonded labile base protons the ^{1}H–^{1}H distances between adjoining stacked base pairs are long compared to the intrabase-pair distances (Tables 11.1 and 11.4). As a consequence, extensive cross relaxation within the base pairs can be expected under the experimental conditions for NOE studies of connectivities between adjoining stacked base pairs.

CHAPTER 12
Resonance Assignments in Nucleic Acids Using Scalar Couplings

This chapter describes those aspects of NMR assignments in nucleic acids that make use of through-bond, scalar spin–spin couplings. These procedures are not dependent on prior knowledge of the molecular conformation, since they rely only on the *presence* of observable spin–spin couplings but *not on quantitation* of the spin–spin coupling constants. They include identification of the ^{1}H spin systems contained in the individual mononucleotides (Table 2.5) as well as sequential assignments via ribose proton–^{31}P couplings (Fig. 11.1). All these studies are preferably done with D_2O solutions, since the scalar coupling networks do not include any of the NMR-observable, labile protons (Table 2.5).

12.1. IDENTIFICATION OF ^{1}H SPIN SYSTEMS

We first consider the spin systems of the protons in the purine and pyrimidine rings (Table 2.5). The isolated, single protons in position 8 of G and in positions 2 and 8 of A give sharp singlet lines. These are best identified in 1D ^{1}H NMR spectra (Figs. 3.7 and 3.8). Their identity can further be confirmed by the fact that the corresponding diagonal peaks are suppressed in DQF-COSY. Since all these singlets have similar chemical shifts (Table 2.6; Figs. 3.7 and 3.8), further distinction must rely on other parameters. Specifically, in A- or B-DNA the 2-protons of adenine have outstandingly long T_1 relaxation times (Weiss et al., 1984b) and at elevated temperature the protons in position 8 may be exchanged with the deuterium of the solvent D_2O, where this exchange is faster for G than for A (Patel et al., 1982a).

The AX spin systems of C and U and the A₃X spin system of T are manifested by cross peaks in otherwise empty regions of the COSY spectrum. For example, in the dodecanucleotide duplex d(CGCGAATTCGCG)₂ the cross peaks of the four AX systems of C and the two A₃X systems of T are well resolved in the spectral regions f and g (Fig. 5.12), respectively, so that the corresponding diagonal positions can readily be identified (see Fig. 8.5A, AX and A₃X systems). If these spectral regions are more crowded in larger compounds, one will want to take advantage of the improved resolution of phase-sensitive COSY or DQF-COSY (Fig. 5.18) and make use of the antiphase cross-peak fine structure (Fig. 5.28) for resolving partially overlapped resonances.

The labile ring protons in nucleic acid duplexes are observed as singlets in 1D ¹H NMR spectra recorded in H₂O (Figs. 3.8 and 3.10). The amino-proton resonances at 6.5 to 9.0 ppm can be distinguished from the resonances of the nonlabile ring protons by their disappearance in D₂O solution, and from those of the imino-proton lines on the basis of the different chemical shift ranges (Table 2.6). A usually quite reliable classification of the imino-proton lines results from the observation that the chemical shifts for A≡T or A≡U are on the low-field side of 13.0 ppm, and those for G≡C at high field from 13.0 ppm (Chou et al., 1983; Robillard and Reid, 1979; Weiss et al., 1984b).

The seven-proton deoxyribose spin systems in DNA typically cover a chemical shift range of approximately 4 ppm (Table 2.6) and have weak spin–spin coupling between 1′H and 2′CH₂, 2′CH₂ and 3′H, and 3′H and 4′H, with the COSY cross peaks well separated from the diagonal (Fig. 12.1). Strong coupling is likely to complicate the spectral analysis for 4′H, 5′H, and 5″H, an exception being 5′H and 5″H at nonphosphorylated 5′ chain ends, which are typically 0.4 ppm to higher field than the other 5′CH₂ lines. As an illustration the complete COSY connectivities (Fig. 12.1) for the four deoxyribose spin systems in the single-strand tetranucleotide d(CTAG) are outlined in the absolute-value COSY spectrum of Figure 12.2. All cross peaks could be identified even in this low-resolution spectrum. Figure 5.12 shows that in a double-helical dodecamer a similar spectrum provided connectivities from 1′H to 2′CH₂ to 3′H and, at least in part, to 4′H (Hare et al., 1983), whereas the connectivities among 4′H, 5′H, and 5″H are strongly overlapped with each other and with the diagonal.

Generally, with a suitable selection of phase-sensitive recordings of COSY, ω₁-scaled COSY, DQF-COSY, RELAYED-COSY, DOUBLE-RELAYED-COSY, MQ spectra, TOCSY, and possibly other experiments in the future (Section 5.5; Figs. 5.18, 5.20, 5.26–5.34, and 12.1), nearly complete deoxyribose spin system identifications including 1′H, 2′H, 2″H, 3′ H, and 4′ H should be attained entirely by scalar through-bond connectivities. In addition, in small molecules the J connectivities bewteen 1′H and the base protons might be observed in experiments such as J-scaled COSY (Gundhi et al., 1985). As an illustration, Figure 12.3 shows the spectral regions d and e (Fig. 5.12) from a phase sensitive, ω₁-scaled DQF-COSY spectrum of

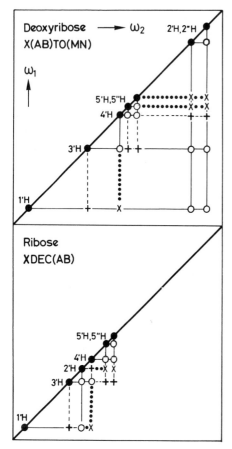

Figure 12.1. COSY, RELAYED-COSY and DOUBLE-RELAYED-COSY connectivity diagrams for the spin systems of nonlabile protons in 2'-deoxy-β-D-ribose and β-D-ribose. Same presentation as Figure 8.5, with ●, ○, +, and × representing diagonal peaks, COSY cross peaks, RELAYED-COSY cross peaks, and DOUBLE-RE-LAYED-COSY cross peaks, respectively. The chemical shifts are approximately to scale, the relative order of nearby resonances is arbitrary.

d(GCATTAATGC)$_2$, which contains the cross peaks between 1'H, 2'H, 2"H, and 3'H (Chazin et al., 1986). For all nucleotides except C10, two cross peaks linking 1'H with 2'CH$_2$ are observed. The fine-structure patterns for the two peaks are characteristically different, which manifests the differences between the coupling constants $^3J_{1'2'}$ and $^3J_{1'2''}$ as well as between $^3J_{2'3'}$ and $^3J_{2''3'}$ (Fig. 5.28). Throughout, a larger coupling prevails for the resonance at higher field. Assuming a 2'C-*endo* ring pucker, this provides an initial indication that the higher-field lines correspond to 2'H (Altona, 1982). For C10 the fine structure of the single peak shows that the 2'H and 2"H resonances have nearly identical chemical shifts. For each nucleotide only a single cross peak between 2'CH$_2$ and 3'H is seen, connecting the 2'H line. This is again compatible with a 2'C-*endo* ring pucker, where $^3J_{2''3'}$ is expected to be of the order 1–2 Hz (Altona, 1982), so that the intensity of the corresponding antiphase COSY cross peaks is very small (Fig. 5.32).

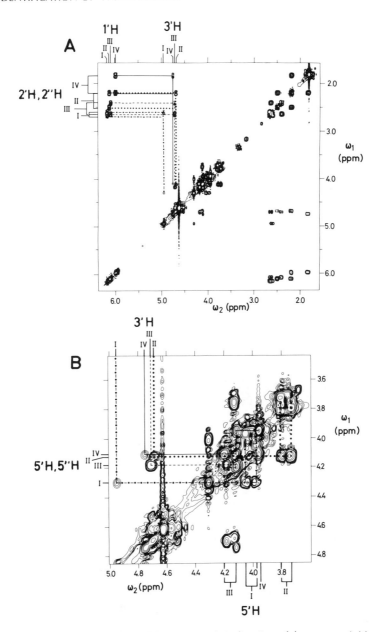

Figure 12.2. ¹H COSY spectrum of d(CTAG) with indication of the connectivities in the deoxyribose spin systems (0.02 *M*, D₂O, pD 8.0, 50°C; 500 MHz; absolute value). The deoxyriboses are arbitrarily denoted I to IV, and the chemical shifts of the individual protons are indicated along the margins. *(A)* Survey plot. *(B)* Expanded plot of the region containing 3′H, 4′H, 5′H, and 5″H. (Reprinted with permission from Pardi et al., *J. Am. Chem. Soc.*, **105**, 1652–1653; © 1983 American Chemical Society).

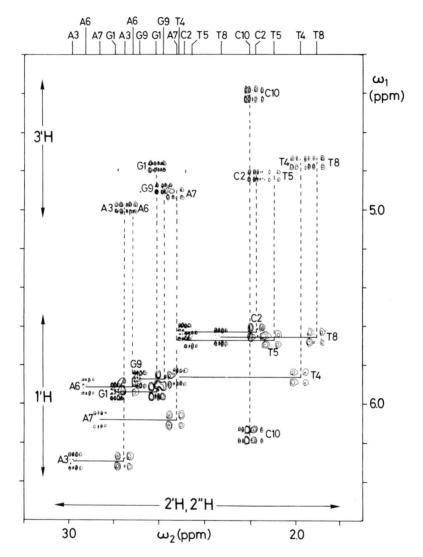

Figure 12.3. Spectral regions d and e (Fig. 5.12) of a phase sensitive, ω_1-scaled ^{1}H DQF-COSY spectrum of d(GCATTAATGC)$_2$ (0.005 M in duplex, D$_2$O, 0.1 M NaCl, 0.05 M phosphate, pD 7.0, 28°C; 500 MHz; scaling factor 0.5; digital resolution 3.9 Hz/point in ω_1, 1.9 Hz/point in ω_2). The chemical shift ranges for 1'H and 3'H are indicated by the arrows on the left, those for 2'H and 2"H at the bottom. Sequence-specific assignments obtained with the techniques of Section 13.2 are indicated for 2'H and 2"H at the top, and for 1'H and 3'H next to the cross peaks in the figure. For each nucleotide the cross peaks 1'H–2'H and 1'H–2"H are connected by a solid horizontal line, and the cross peaks 1'H–2'H and 2'H–3'H by a broken vertical line (from Chazin et al., 1986).

The six-proton ribose spin systems in RNA cover a chemical shift range of only 2 ppm (Table 2.6). Except for the well separated 1'H line, all resonances are likely to be connected by strong coupling, and even for relatively small RNA fragments the spectral region of the resonances 2'H–5'H is usually very crowded (Clore et al., 1984, Lankhorst et al., 1985). With relayed coherence-transfer techniques it should be possible to identify the chemical shifts of 2'H and 3'H, possibly even 4'H, in ribose spin systems that give rise to a resolved 1'H line (Fig. 12.1). The practical use of these spin system identifications for further studies may be limited, however, by chemical shift degeneracies between different ribose moieties in the RNA fragment.

In conclusion it should be emphasized that careful, complete spin system identifications by scalar connectivities are indispensable as a basis for reliable sequential resonance assignments (Section 12.2, Chapter 13).

12.2. SEQUENTIAL ASSIGNMENTS WITH ^{1}H–^{31}P COUPLINGS

The polynucleotide backbone contains a continuous network of vicinal spin–spin coupling constants: $^3J_{4'3'}$ (i), $^3J_{3'^{31}P}$ (i), $^3J_{^{31}P5'}$ $(i + 1)$, $^3J_{5'4'}$ $(i + 1)$, $^3J_{4'3'}(i + 1)$, . . . (Fig. 11.1). Both ^{1}H and ^{31}P are abundant nuclei with high NMR sensitivity (Table 2.1), so that a favorable situation prevails for experimental observation of the heteronuclear couplings. Figure 12.4 presents a survey of the procedures for sequential assignments using ^{1}H–^{1}H and ^{1}H–^{31}P scalar coupling connectivities.

The example used to illustrate the assignment strategy is d(CTAG), which adopted a flexible single-strand conformation under the conditions of these experiments. In the first step (Fig. 12.4) the 1'H resonances were labeled I–IV in the order of decreasing chemical shift (Fig. 12.2A), and correspondingly the ^{31}P resonances were labeled 1–3 (Fig. 12.5A). The second step consisted of the identification of the four deoxyribose spin systems with homonuclear ^{1}H COSY, which started from the 1' protons I–IV (Fig. 12.2A and B). In the third step the scalar coupling connectivities between the ^{1}H spin systems I–IV and the ^{31}P lines 1–3 were identified using ^{1}H–^{31}P COSY. In Figure 12.5B to D, cross sections along the ^{1}H chemical-shift axis were plotted at the ω_2 positions of the three ^{31}P lines. From comparison of the resonance positions in these cross sections with the previously determined chemical shifts for the ribose spin systems I–IV, the ^{1}H spin systems coupled with the phosphates 1, 2, and 3 were identified as listed in Figure 12.4. Finally, the three ribose–phosphate–ribose fragments thus obtained were assembled in a linear sequence; clearly, from the mutual overlaps of these fragments the assembly shown at the bottom of Figure 12.4 is unambiguous. In this special case of a tetranucleotide with four different bases, complete sequence-specific assignments for all ^{1}H and ^{31}P resonances were thus obtained (Pardi et al, 1983a).

ASSIGNMENT FOR d(CTAG)

① 1'H resonances   I to IV
   31P resonances   1 to 3

② J(1H,1H) ⟶ 1H spin systems I to IV

③ J(1H,31P) ⟶ II3'—1—5'IV
                 I3' —2—5'III
                 IV3'—3—5'I

④ d(CTAG) ∿ II–1–IV–3–I–2–III

Figure 12.4. Survey of the experimental steps in the sequential resonance assignments for d(CTAG) using 1H–31P heteronuclear couplings.

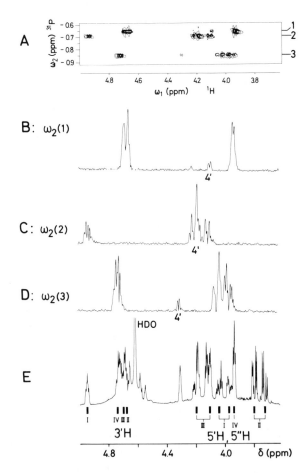

Figure 12.5. Heteronuclear 1H–31P COSY spectrum of d(CTAG) (0.009 M, D$_2$O, pD 8.0, 50°C; 300 MHz 1H; absolute value). *(A)* Contour plot. *(B)–(D)* Cross sections along the 1H-frequency axis at the ω_2 frequencies of the 31P resonances 1–3. *(E)* 1D 1H NMR spectrum with identification of the 3'H-, 5'H-, and 5"H-chemical shifts. (Reprinted with permission from Pardi et al., *J. Am. Chem. Soc.*, **105**, 1652–1653; © 1983 American Chemical Society.)

For use of this technique with larger DNA fragments, it is important that connectivities between ³¹P and 4'H can be established either by observation of direct coupling or by relay techniques, since the protons 5' and 5" cannot usually be identified. In Figure 12.5*B* to *D*, the 4'H-chemical shifts are indicated, and it is readily apparent that direct couplings ³¹P–4'H are also manifested in this COSY experiment. Complete assignments for the backbone of the hexanucleotide-duplex d(ATGCAT)₂ were obtained using the ³¹P–4'H couplings (Marion and Lancelot, 1984). A further limiting factor in larger molecules is overlap of the ³¹P resonance lines. This potential difficulty can be overcome with the use of ³¹P relayed ¹H–¹H COSY (Delsuc et al., 1984; Neuhaus et al., 1984b), which was used for complete backbone assignments in another hexanucleotide duplex d(GCATGC)₂ (Frey et al., 1985a).

An interesting extension of the use of resonance assignments with ³¹P–¹H couplings is by combination with an ¹⁷O isotope-labeling technique. This relies on the fact that the ³¹P resonance in ¹⁷O-labeled phosphate groups is broadened beyond detection (Joseph and Bolton, 1984; Petersheim et al., 1984). Figure 12.6 shows the ³¹P spectra of a series of selectively ¹⁷O-labeled analogs of the octanucleotide-duplex d(GGAATTCC)₂. The isotope enrichment was close to 50%, so that the resonance intensity of the labeled phosphates is approximately halved. The series of experiments in Figure 12.6 required a considerable amount of synthetic work, but clearly resulted in unambiguous, sequence-specific ³¹P resonance assignments (Connolly and Eckstein, 1984). Similar work with different oligonucleotides was extended by heteronuclear ³¹P–¹H COSY to include sequence-specific assignments for the deoxyribose protons (Gorenstein et al., 1984; Lai et al., 1984). Potentially interesting applications for this combination of isotope labeling and heteronuclear correlation spectroscopy include site specific, partial reso-

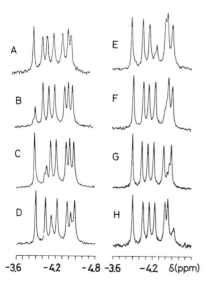

Figure 12.6. 1D ³¹P NMR spectra of selectively ¹⁷O-labeled d(GGAATTCC)₂ (D₂O, 0.02 *M* Hepes, pD 8.0, 0.02 *M* EDTA, 0.05 *M* NaCl, 10°C; 81 MHz). (A) Unlabeled. (B)–(H) ¹⁷O-labeled phosphate in the 3' position of C7, G1, G2, T6, A3, T5, and A4, respectively. (Reprinted with permission from Connolly and Eckstein, *Biochemistry,* **23,** 5523–5527; © 1984 American Chemical Society.)

nance assignments, which might possibly be used with larger molecules, or with RNA fragments.

Assignments with the techniques described in this section are restricted to the ribose protons and the ^{31}P resonances of the polynucleotide backbone. In duplex conformations, the base protons can be assigned via $^1H-^1H$ NOE's with the methods described in Chapter 13. For possible other nonflexible forms, for example, in complexes of oligonucleotide single strands with polypeptides or other molecules, the data on intranucleotide distances in Figures 11.5 and 11.6 imply that attachment of the base protons by $^1H-^1H$ NOE's should in general also be feasible. In small flexible oligonucleotides the J couplings between $1'H$ and the base protons can possibly be used for extending the assignments to the bases (Gundhi et al., 1985).

CHAPTER 13
Nucleic Acid Conformation, ¹H–¹H Overhauser Effects, and Sequence-Specific Resonance Assignments

This chapter includes a description of sequential resonance assignments with ¹H–¹H NOE's, which is related to the sequential assignments in proteins (Chapter 8). There are also, however, important differences relative to the situation in proteins. Thus, since each mononucleotide contains two or three separate ¹H spin systems, resonance assignments within the individual building blocks of nucleic acids depend more extensively on NOE connectivities than within the common amino acids. Furthermore, in contrast to the situation in polypeptide chains, all sequential ¹H–¹H distances in polynucleotides can independently adopt values that are too long for observation by NOE's. A somewhat different assignment strategy is therefore required. Most important, before working on the assignments, the conformation type of the polynucleotide must be established with the use of suitable NMR experiments and possibly other spectroscopic techniques. Once it has thus been determined that the molecular conformation resembles one of the standard helical forms, the distance relations in Tables 11.1 to 11.4 and Figures 11.4 to 11.7 can provide a basis for sequential assignments by NOE's. For A- and B-type helical structures, the assignments for the nonlabile protons in the individual polynucleotide strands and for the hydrogen-bonded labile protons in the base pairs are then best obtained from separate experiments recorded, respectively, in D_2O and in H_2O.

In Chapter 11 we found that short ¹H–¹H distances in nucleic acid helices occur exclusively within individual mononucleotides, between sequentially neighboring nucleotides and between directly adjacent, stacked base pairs. Sequential resonance assignments and studies of the spatial structure will therefore largely have to rely on observation of the same NOE's. Since

233

structure determination and sequential assignments are thus in several ways strongly interdependent, the two subjects are both treated in this chapter.

13.1. CHARACTERIZATION OF THE CONFORMATION TYPE

Over the years several spectroscopic experiments have been shown to provide quite reliable criteria for differentiating between different nucleic acid conformations. These are quick and efficient analytical tools, since they rely on measurement of a small number of parameters, which are correlated in an empirical fashion with the molecular conformation.

As discussed in Sections 3.2 and 3.3 the ¹H chemical shifts in nucleic acids are markedly dependent on conformation (Figs. 3.7 and 3.12; Giessner-Prettre and Pullman, 1970; Giessner-Prettre, 1984; Robillard and Reid, 1979; Tran-Dinh et al., 1983), so that chemical shift measurements give an initial indication for the presence of a stacked, "structured," or flexible, "unstructured" polynucleotide chain. More definite information is obtained if the chemical shift differences are used for observation of conformational transitions upon changes of the solution conditions. As an illustration Figure 13.1 shows the effects of *thermal melting* on the chemical shifts of the dodecamer duplex d(CGCGAATTCGCG)₂. The chemical shifts in the stacked conformation are up to 0.5 ppm to higher field. During the transition the two forms are in rapid exchange, so that a continuous melting curve is obtained (Patel et al., 1982a; Melema et al., 1984a).

Observation of the hydrogen-bonded imino protons in Watson–Crick base pairs, which have characteristic chemical shifts at low field (Figs. 3.8–3.10),

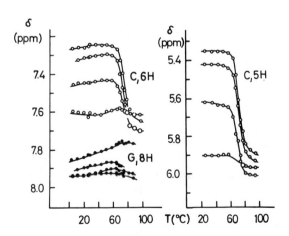

Figure 13.1. Melting of the DNA duplex d(CGCGAATTCGCG)₂. The chemical shifts of the nonlabile ring protons of G and C are plotted versus temperature (D₂O, 0.002 *M* EDTA, 0.1 *M* phosphate, pD 7.7) (from Patel et al., 1982a; © 1982 by the AAAS).

is direct evidence for the formation of double-helical structures. A comparison of the number of imino-proton NMR lines originating from A=T and G≡C base pairs with the corresponding number expected either from the primary structure of short nucleic acid fragments or from spatial molecular models of more complex molecules, such as tRNA, usually provides clear indications of the hydrogen-bonding network in the molecule studied.

Three methods for distinguishing between right-handed and left-handed DNA helices are circular dichroism (Pohl and Jovin, 1972), ^{31}P NMR chemical shifts (Patel et al., 1982b; Holak et al., 1984), and observation of the extremely short ^{1}H–^{1}H distance $d_i(8;1')$ for purine nucleotides with syn conformation in Z-DNA (Table 11.5) (Patel et al., 1982b; Feigon et al., 1983). (Additional techniques include ultraviolet, infrared, and Raman spectroscopy.) Figure 13.2 illustrates the change of the ^{31}P NMR spectrum of poly(dG–dC) during a salt-induced transition from the B to the Z conformation. In B-DNA the chemical shift dispersion of the ^{31}P resonances is small, giving rise to a single, composite line. In contrast, for Z-DNA the conformational nonequivalence of the two phosphate groups in the dinucleotide repeat (Table 11.5) is reflected in distinctly different ^{31}P chemical shifts. (In Fig. 13.2 the line broadening at high salt concentration was attributed to the presence of different, minor conformations.) The short distance $d_i(8;1')$ resulting from the syn conformation of purine nucleotides in Z-DNA (Fig. 13.3) is unique since the anti conformation prevails for all nucleotides in A- or B-DNA (Fig. 11.5). Therefore, the corresponding strong NOE is an indication for the presence of the Z conformation.

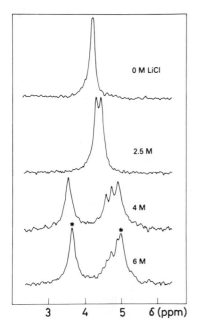

Figure 13.2. 1D ^{31}P NMR spectra of poly(dG–dC) at variable concentrations of LiCl (D$_2$O, 0.001 M EDTA, 0.01 M sodium cacodylate, pD 7.2; 81 MHz; proton noise decoupled; δ relative to trimethylphosphate). The asterisks indicate the ^{31}P resonance positions characteristic of Z-DNA (from Patel et al., 1982b).

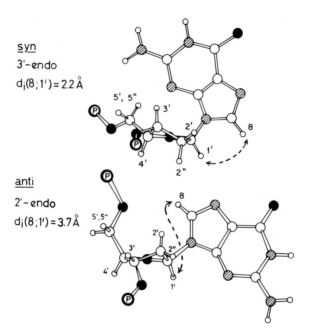

syn
3'-endo
$d_i(8;1') = 2.2$ Å

anti
2'-endo
$d_i(8;1') = 3.7$ Å

Figure 13.3. Pictorial presentation of the dependence of $d_i(8;1')$ on the glycosidic torsion angle χ (Fig. 11.5). The two conformations for G prevail in the Z form of d(CGCGCG)$_2$ (top) and in B-DNA (bottom) (from Patel et al., 1982b).

13.2. SEQUENTIAL ASSIGNMENTS OF NONLABILE PROTONS IN B-DNA USING ^1H–^1H OVERHAUSER EFFECTS

Right-handed, double-helical DNA in aqueous solution can quite safely be assumed to be of the B-DNA type (Dickerson, 1983; Saenger, 1984). Starting with this assumption, different authors proposed procedures for sequential assignments of the nonlabile protons with the use of NOESY (Broido et al., 1984; Clore and Gronenborn, 1985b; Feigon et al., 1983; Frechet et al., 1983; Haasnoot et al., 1983b; Hare et al., 1983; Hosur et al., 1985; Scheek et al., 1983; 1984; Weiss et al., 1984a) or TOE difference spectra (e.g., Reid et al., 1983; Sanderson et al., 1983). Notwithstanding differences in details, all the proposed strategies rely on the sequential connectivities in Figure 13.4. While it may appear from this illustration that the entire assignment problem could be solved by NOESY alone, it must be reemphasized that the assignments can be placed on more solid ground by exhaustive use of scalar coupling connectivities for spin system identification prior to the sequential assignments.

The intranucleotide and sequential NOE connectivities in Figure 13.4 are a selection from Figures 11.5 and 11.6 and Table 11.2. In the 5'-to-3' direction the figure outlines two continuous networks of short ^1H–^1H distances,

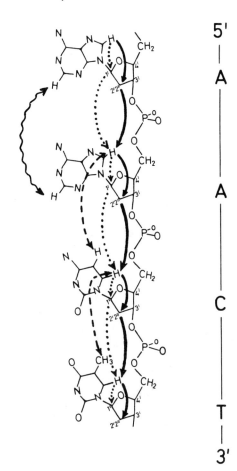

5'
|
A
|
A
|
C
|
T
|
3'

Figure 13.4. Pictorial presentation of the short ^{1}H–^{1}H distances $d_i(6,8;2')$ and $d_s(2'';6,8)$ (solid arrows), $d_i(6,8;1')$ and $d_s(1';6,8)$ (dotted arrows), $d_s(6,8;5,M)$ (broken arrows) and $d_s(2;2)$ (wavy arrow), which are used for sequential assignments of the nonlabile protons in B-DNA.

namely, $d_i(6,8;2')$ followed by $d_s(2'';6,8)$ and $d_i(6,8;1')$ followed by $d_s(1';6,8)$. Short distances between base protons of neighboring nucleotides, $d_s(6,8;5,M)$, prevail for pyrimidine–pyrimidine and purine–pyrimidine dinucleotide segments [for the notation used see Eq. (11.8)]. The short distance $d_s(2;2)$ between neighboring adenines is discussed at the end of this section. The implications of Figure 13.4 must be evaluated in light of three additional major factors affecting the assignment procedure, namely, the identification of starting locations for the sequential assignments, the influence of spin diffusion, and limitations that may arise from incomplete spectral resolution of certain proton types.

Identification of the chain-terminal nucleotides as reference locations for the sequential assignments can follow from different, independent criteria. If the terminal deoxyriboses are not phosphorylated, they can be identified from the high field shifts of their 5'CH$_2$ or 3'H, and from the absence of ^{31}P couplings with these protons. Since the short sequential distances from deoxyribose to base protons are in the direction 5′ to 3′ (Fig. 13.4), the 5′-terminal base and the 3′-terminal deoxyribose are unique in exhibiting only intranucleotide NOE's, which can be used for their identification. Nonterminal sequence locations can be identified from similar considerations to those used for proteins, namely, by matching the sequence of oligonucleotide segments assembled with sequential NOE connectivities against corresponding segments in the primary structure, which is known from the chemical synthesis of the DNA or from sequence analysis. (Quite naturally, since there are only four different nucleotides, the probability that identical di- and trinucleotide sequences occur repeatedly in the same strand is intrinsically higher than for di- and tripeptides in proteins (Table 8.2), but then the chain length of the species studied is usually short even compared to small proteins.) Obviously, if the DNA fragments are obtained by chemical synthesis, specified nucleotides could also be labeled by isotope enrichment or chemical modification to enable NMR assignments (Section 4.2).

The effects of spin diffusion in NOE experiments used for sequential resonance assignments can be assessed from Figure 11.4 and Table 11.1, which show that in B-DNA the following ^{1}H–^{1}H distances are comparable to or shorter than the connectivities in Figure 13.4: $d_i(5;6)$ in C, $d_i(M;6)$ in T, $d_i(2′;2″)$, $d_i(1′;2″)$, $d_i(1′;2′)$, $d_i(2′;3′)$, $d_i(2″;3′)$, $d_i(3′;4′)$, and possibly $d_i(1′;4′)$. The shortest distance is $d_i(2′;2″)$, so that these two geminal protons are thoroughly short circuited. Spin diffusion will further be effective around the entire deoxyribose ring, and between the base protons of C and T. An illustration is provided by Figure 13.5, which shows the buildup of intraresidue and sequential NOE's on individual deoxyribose protons in the decamer duplex d(GCATTAATGC)$_2$. The cross section showing the NOESY cross peaks with 6H of thymine-4 is representative for the observations made for all dinucleotide segments in this duplex (Chazin et al., 1986). After a mixing time of 30 ms only the peaks corresponding to the shortest base proton–ribose proton distances, $d_i(6,8;2′)$ and $d_s(2″;6,8)$, had sizable intensity. In measurements with longer mixing times, cross peaks corresponding to intranucleotide and sequential connectivities with 1′H, 2′H, and 2″H were observed, and there were also quite large NOE intensities for $d_i(6,8;3′)$ and $d_s(3′;6,8)$, even though in the standard B form of DNA these two distances are of the order of 4.5 Å (Fig. 11.6 and Table 11.2) and thus at the limit for direct NOE detection.

Overall there is little danger that spin diffusion would falsify the resonance assignments, since cross relaxation will be most effective around the deoxyribose rings within the individual nucleotides, and possible additional sequential cross relaxation would be in the direction of the assignment pathways in Figure 13.4 (Hare et al., 1983). In situations where certain 1′H, 2′H, or 2″H resonances are not resolved, one may even wish to promote spin diffusion in order to follow the sequential connectivities via 3′H or even 4′H. On the other hand, when using mixing times longer than approximately 50

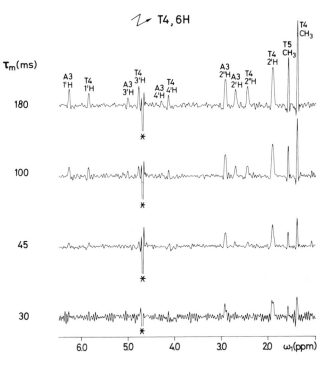

Figure 13.5. NOE buildup and spin diffusion in d(GCATTAATGC)$_2$. Cross sections along ω_1 at ω_2 = 7.16 ppm (this corresponds to the chemical shift of 6H in T4) from NOESY spectra recorded with the mixing times indicated in the figure are shown (same solution as Fig. 3.8A; 500 MHz; absorption mode, digital resolution 3.9 Hz/point in ω_1 and 1.9 Hz/point in ω_2). Resonance assignments are indicated in the top trace. The star identifies an experimental artifact (from Chazin et al., 1986).

ms, the NOE's between base protons and different deoxyribose protons should not be taken as independent evidence for the resonance assignments, since they will all be short circuited by spin diffusion.

As a first practical example, Figure 13.6A presents the connectivities between TCH$_3$ and 6H or 8H in region a, and between 2'H and 2"H, and 6H or 8H in region g (Fig. 6.9) of a NOESY spectrum of d(GCATTAATGC)$_2$ recorded with a long mixing time allowing extensive spin diffusion. To facilitate access to the information contained in this figure, the sequence-specific assignments obtained as the final result are indicated, and the intraresidue and sequential cross peaks are identified. In the following, the arguments leading to these assignments are presented.

For the three TCH$_3$ resonances, the strong intrabase NOE to 6H was identified from comparison with COSY (see Fig. 5.12). In Figure 13.6A these cross peaks are connected by horizontal lines with the sequential connectivities. Sequential NOE's of TCH$_3$ are expected with 6H or 8H of the preceding

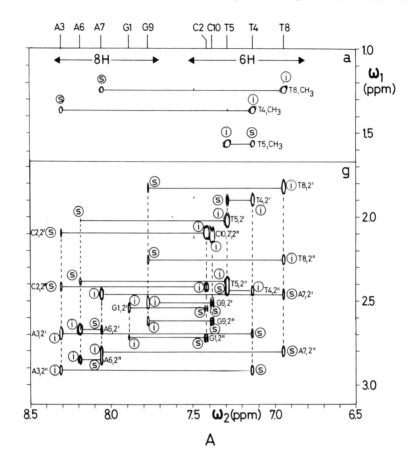

A

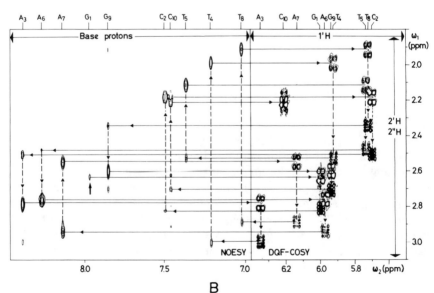

B

nucleotide (Table 11.2), so that 6H of the first T in a TT dinucleotide segment should have NOE's with two methyl groups. This was observed for the 6H resonance at 7.16 ppm, which was therefore assigned to T4 in the decamer sequence. Via the sequential NOE connectivities with CH_3 and 6H of T4, the chemical shifts of 8H in A3 and 6H in T5 were then identified. From comparison with the decamer sequence, the remaining methyl resonance had to come from T8, and from the NOESY cross peaks with this resonance the chemical shifts of 8H in A7 and 6H in T8 were identified. As a result of the spectral analysis in region a, sequence-specific assignments were thus obtained for 6H or 8H of five nucleotides. These were subsequently used as starting points for the sequential assignments in the spectral region g.

Since the NOESY spectrum was recorded with a mixing time of 300 ms, spin diffusion caused similar spin temperatures for the individual protons within each $2'CH_2$ group (Fig. 13.5). For 6H or 8H of the nucleotides 2–10 one therefore expects a pattern of four cross peaks with 2'H and 2"H. Conversely, for each 2'H and 2"H of the residues 1–9 two cross peaks with 6H or 8H are anticipated. For 8 base protons, complete sets of four cross peaks were observed; in the figure these are connected by vertical, broken lines. There were also a 6H resonance with three cross peaks at 7.39 ppm and an 8H resonance with two cross peaks at 7.91 ppm. The latter was attributed to the 5'-terminal G1. For eighteen 2'H and 2"H resonances complete sets of two cross peaks were observed; these are connected by horizontal, solid lines. An additional resonance at 2.16 ppm showed a single cross peak, which relates this line with the 6H resonance that is connected with only three $2'CH_2$ resonances. An explanation for these incomplete cross-peak patterns (which was subsequently confirmed by the sequential assignments) is that 2'H and 2"H of C10 have identical chemical shifts at 2.16 ppm (Fig.

Figure 13.6. Sequential resonance assignments in d(GCATTAATGC)$_2$ using NOESY connectivities between 2'H, 2"H, or TCH$_3$, and 6H or 8H. *(A)* Regions a and g (Fig. 6.9) of a NOESY spectrum recorded with a mixing time of 300 ms (same solution as Fig. 3.8*A*; 500 MHz; absorption mode; digital resolution 3.9 Hz/point in ω_1, 1.9 Hz/point in ω_2). Sequence-specific assignments for 6H or 8H are indicated at the top, for TCH$_3$, 2'H, and 2"H near the cross peaks in the spectrum. The separate identifications for 2'H and 2"H resulted from the fine structure of the COSY cross peaks in Figure 12.3. The letters i and s identify intraresidue and sequential connectivities, respectively. *(B)* NOESY-COSY connectivity diagram for sequential assignments via NOE's between base protons and 2'C protons using NOESY with a mixing time of 60 ms [otherwise same conditions as *(A)*] and DQF-COSY recorded interleaved with NOESY. In the NOESY spectrum, successive cross peaks $d_s(2";6,8)$ and $d_i(6,8;2')$ are connected by dashed vertical lines. In the COSY spectrum the cross peaks 1'H–2'H and 1'H–2"H of each nucleotide are connected by dashed vertical lines. Sequence-specific assignments for the base protons and the 1' protons are indicated at the top. Horizontal lines indicate the NOE connectivities $d_i(6,8;2')$ and $d_s(2";6,8)$. Following the arrows the sequential assignment pathway can be followed from G1 to C10 (adapted from Chazin et al., 1986).

12.3) and that the line at 7.39 ppm corresponds to 6H of C10. Starting from any of the resonances that were so far individually assigned, namely, 8H of G1, A3 and A7, 6H of T4, T5, and T8, and 2'H and 2"H of G1, continuous assignment pathways i–s–i–s– ·· can be traced by following alternatively a solid and a broken line in Figure 13.6, region g. Since each intranucleotide and sequential connectivity can be selected either via 2'H or 2"H, a variety of different geometric patterns can be composed in the figure, all of which represent a valid assignment pathway.

As in Figure 13.6*A*, the principal assignment pathway (solid arrows in Fig. 13.4) is also used in Fig. 13.6*B* for an alternative presentation of the sequential assignments in the form of a NOESY-COSY connectivity diagram (Chazin et al., 1986), which is patterned after corresponding plots for sequential assignments in proteins (Figs. 8.13 and 8.14; Wagner et al., 1981). One component of this diagram is the region in a DQF-COSY spectrum which contains the cross peaks between 1'H and 2'CH$_2$. Broken vertical lines in this spectrum connect the cross peaks 1'H–2'H and 1'H–2"H for each nucleotide, with the 1'H resonance identified at the top of the figure. A second component is region g (Fig. 6.9) of a NOESY spectrum recorded with conditions identical to the COSY experiment, and with a sufficiently short mixing time so that only the connectivities $d_i(6,8;2')$ and $d_s(2";6,8)$ are observed (Fig. 10.5). Thus there is one intraresidue and one sequential connectivity for each base (except that only the intranucleotide connectivity prevails for the 5' terminus). Broken vertical lines link the two NOESY cross peaks of each base proton resonance, with the base resonance identified at the top of the figure.

The 5'-terminal base proton is a convenient starting point for an outline of the assignment pathway in the NOESY-COSY connectivity diagram. From the $d_i(8;2')$ NOESY cross peak of G1 (arrow in Fig. 13.6*B*) a horizontal line leads to the COSY cross peak 1'H–2'H of G1, which is in turn connected to the COSY cross peak 1'H–2"H of G1 by a vertical broken line. From there, a horizontal line leads to the sequential NOESY connectivity $d_s(2";6)$, thereby linking G1 to the adjacent nucleotide C2. A vertical line in the NOESY spectrum connects the sequential cross peak with the $d_i(6;2')$ cross peak of C2, and from there the assignment pathway leads back into the COSY spectrum to the 1'H–2'H cross peak of C2. Following the arrows in Figure 13.6*B* the assignments can be traced all the way to C10. Each sequential assignment step corresponds to a complete cycle in the NOESY-COSY connectivity diagram. These cycles are either clockwise or anticlockwise, depending on the chemical shifts of the sequentially neighboring nucleotides.

Figure 13.6*B* emphasizes the contributions from scalar coupling connectivities to the sequential assignment pathway. The NOESY connectivities used are the direct NOE's via two unique, outstandingly short intranucleotide and sequential ^{1}H–^{1}H distances (Fig. 11.6 and Table 11.2). The use of very short mixing times for NOESY has the additional advantage for work with larger, more complex DNA fragments that the spectra contain a smaller number of cross peaks and are therefore easier to interpret (compare Figs. 13.6A and 13.6B).

Figure 13.7 illustrates a second assignment pathway via $d_i(6,8;1')$ and $d_s(1';6,8)$ (dotted arrows in Fig. 13.4) for the duplex d(CGCGAATTCGCG)$_2$ (Hare et al., 1983). As a starting location for the sequential assignments, the base protons for the TT segment in positions 7 and 8 were identified as in region a of Figure 13.6A. For each 6H or 8H line except 6H of C1, one expects two cross peaks with 1'H resonances, since the NOESY spectrum was recorded with a long mixing time of 300 ms. Such patterns, which are connected by vertical lines in the figure, were observed for five 6H positions. The single remaining 6H resonance had only one NOE with 1'H and could therefore be attributed to C1. For three 8H positions the expected two cross peaks were observed, for one position there was only one cross peak, and at the location of two overlapping 8H lines three cross peaks were seen. For all 1'H lines except G12 one anticipates two cross peaks with 6H or 8H. These two-peak patterns, which are connected by horizontal lines, were observed for nine 1'H positions. The remaining three 1'H resonances had only a single cross peak each, and thus had to correspond to the 3' terminal G12, to one of

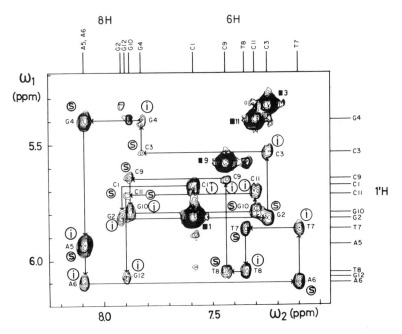

Figure 13.7. Intraresidue (i) and sequential (s) connectivities between 1'H and 6H or 8H in the region t of the NOESY spectrum of d(CGCGAATTCGCG)$_2$ in Figure 6.9. Sequence-specific assignments are indicated on the right for 1'H and at the top for 6H and 8H. Four strong cross peaks between 6H and 5H of C are identified by filled squares (from Hare et al., 1983).

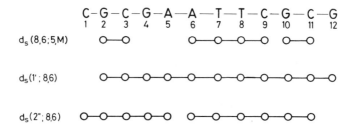

Figure 13.8. Survey of the sequential NOE's used for the resonance assignments in d(CGCGAATTCGCG)$_2$ (from Hare et al., 1983).

the residues with overlapping 8H lines, and to another, unknown nucleotide. Starting from 6H of T8 and following alternatively a horizontal and a vertical line, sequential assignments for the residues 8–12 were obtained as indicated by the arrows in Figure 13.7. In the opposite direction a continuous train of connectivities was found for the segment of residues 2–8. The missing connectivity thus had to be between C1 and G2, as was subsequently confirmed by analysis of the connectivities with 2'H and 2"H in the spectral region g (Fig. 13.8; Hare et al., 1983).

A synoptic presentation of sequential connectivities is afforded by the scheme in Figure 13.8 (Hare et al., 1983). Circles connected by a line indicate that the connectivity indicated on the left was observed between the two corresponding nucleotides. When reading such diagrams one should keep in mind that because of spin diffusion, $d_s(1';6,8)$ and $d_s(2";6,8)$ would be independent assignment pathways only if they resulted from NOESY experiments recorded with very short mixing times (Fig. 10.5). Table 13.1 lists the chemical shifts of individually assigned protons in the DNA duplex d(CGCGAATTCGCG)$_2$, which illustrates in a different way the results obtained with these assignment procedures.

TABLE 13.1. Chemical Shifts of the Nonlabile Protons of the First Five Residues in d(CGCGAATTCGCG)$_2$[a]

Residue	Chemical Shifts (37°C, pD 7.0)							
	8H	6H	5H	1'H	2'H	2"H	3'H	4'H
C1		7.59	5.84	5.71	1.89	2.35	4.65	4.15
G2	7.92			5.84	2.59	2.68	4.91	4.31
C3		7.24	5.35	5.55	1.81	2.22	4.78	4.17
G4	7.83			5.42	2.63	2.74	4.98	4.29
A5	8.09			5.96	2.65	2.90	5.05	4.44

[a] In parts per million (from Hare et al., 1983).

Usually the protons in position 2 of adenine are assigned from NOE connectivities with labile protons (Section 13.3). However, for DNA segments containing only A and T, a continuous assignment pathway for A2H can also be obtained from experiments in D_2O. This relies on the short sequential distance $d_s(2;2)$ (wavy arrow in Fig. 13.4) and on short interstrand A2H–A2H distances in B-DNA$[d_{ps}(A2+,2;A1-,2)$ in Table 11.4]. Figure 13.9 shows such assignments for the duplex d(CGTTATAATGCG) · d(CGCATTATAACG), which contains a stack of seven A=T base pairs. A sequential NOE from A22 to A21 (see legend to Fig. 13.9 for the numeration used) is followed by an interstrand connectivity from A21 to A5. The cross peak from A5 to A19 is not resolved in this absolute-value spectrum, but the following sequence of interstrand and sequential NOE's from A19 to A7 and from A7 to A8 is clearly manifested. The chemical shifts of A8 and A16 are again too close for the cross peak corresponding to the interstrand NOE to be resolved.

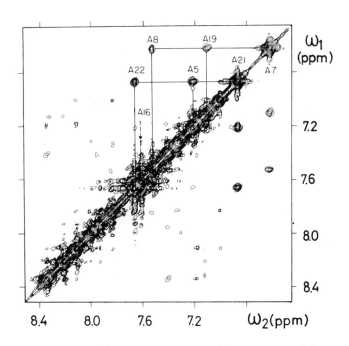

Figure 13.9. Assignment of the A2H resonances of the consensus Pribnow promoter sequence, d(CGTTATAATGCG) · d(CGCATTATAACG), in the region v (Fig. 6.9) of a NOESY spectrum (D_2O, 0.1 M NaCl, 0.01 M phosphate, pD 7.0, 37°C; 500 MHz; τ_m = 300 ms; absolute value). In the direction 5'–3' the nucleotides of the + strand are numbered 1–12, those of the − strand 13–24. The ω_2 positions for A2H are indicated in the figure. (Reprinted with permission from Wemmer et al., *Biochemistry*, **23**, 2262–2268; © 1984 American Chemical Society.)

13.3. ASSIGNMENT OF LABILE PROTONS IN B-DNA USING ¹H–¹H OVERHAUSER EFFECTS

An assignment scheme for labile protons in B-DNA (Boelens et al., 1985; Chou et al., 1983; Gronenborn et al., 1984; Ulrich et al., 1983; Weiss et al., 1984b) can be derived from Tables 11.1, 11.3, and 11.4 (Fig. 13.10). The major assignment pathway is via the NOE's between the hydrogen-bonded imino protons of adjacent stacked base pairs, which are in all cases separated by less than 4.3 Å. For A=T, there are additional short distances with A2H. Since numerous ¹H–¹H distances within the base pairs are much shorter than the distances between protons in adjacent base pairs (Tables 11.1 and 11.4), there is pronounced spin diffusion within the base pairs (Fig. 13.10). The resulting NOE's may be used to identify the base pair types, and if the imino-proton resonances are not resolved they can provide an alternative assignment pathway. The scheme of Figure 13.10 does not include direct NOE's with amino protons in adjacent base pairs, which might also be used in future, refined assignment procedures (Tables 11.3 and 11.4).

The practice of assigning the labile protons in DNA fragments requires that NOE measurements are recorded in H_2O solution with solvent suppression by a semiselective observation pulse (Fig. 5.2). It may also be necessary to work at low temperature, near 0°C, to obtain the desired duplex form and especially to slow down the proton exchange with the solvent. (As an additional difficulty, this causes broadening of the NMR lines.) Up to now 1D TOE difference spectroscopy has mostly been used, but it seems clear that in the future the preferred technique will be NOESY. In the following, the use of both techniques is briefly illustrated.

In Figure 13.11 the TOE difference technique was applied for the heptadecamer duplex d(TACCACTGGCGGTGATA) · d(TATCACCGCCAGTG-GTA) (Weiss et al., 1984b). The A=T imino-proton resonances were labeled $\alpha_1-\alpha_8$, and those of the G≡C base pairs $\gamma_1-\gamma_9$ in order of decreasing chemical shifts (Fig. 13.11A). The TOE difference spectrum obtained with irradiation on α_5 (Fig. 13.11B) shows that α_5 corresponds to an A=T base pair that is sandwiched between two G≡C base pairs, γ_6 or γ_7 and γ_9. In contrast to the sequential connectivities between ribose and base protons (Table 11.2) the short distances between adjacent imino protons do not have a defined

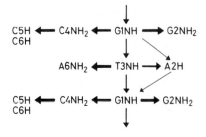

Figure 13.10. Assignment pathway for the labile base protons and A2H in three neighboring base pairs G≡C, A=T, and G≡C in B-DNA. The thickness of the arrows reflects approximately the inverse sixth power of the distance between the protons.

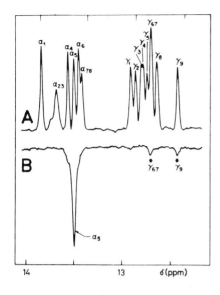

Figure 13.11. Normal 1D ^{1}H NMR spectrum (A) and 1D TOE difference spectrum (B) of the λ-operator site $O_L 1$, d(TACCACTGGCG-GTGATA) · d(TATCACCGCCAGTGGTA) (H$_2$O, 0.001 M EDTA, 0.001 M NaN$_3$, 0.01 M Tris-HCl, pH 7.4, 10°C; 498 MHz; $\tau = 1$ s). The imino-proton lines of the A═T base pairs are labeled $\alpha_1 - \alpha_8$, those of G≡C γ_1 to γ_9 (from Weiss et al., 1984b).

direction, so that the result of Figure 13.11*B* is either $5'-\gamma_{6,7}-\alpha_5-\gamma_9-3'$ or $5'-\gamma_9-\alpha_5-\gamma_{6,7}-3'$. The polarity must therefore be determined by comparison of the segments of stacked base pairs connected via NOE's with the sequence of the compound studied. For this it is important that continuous assignments over several base pairs can be obtained. In Figure 13.11 only the imino-proton resonances are shown, and these were attributed to A═T or G≡C on the basis of the chemical shifts (Section 3.2). Between 6 and 9 ppm the TOE difference spectra may further contain sharp peaks corresponding to A2H, and broader resonances from the amino protons of A, G, and C, which can provide additional evidence for the identification of the base pair types (Chou et al., 1983, 1984; Gronenborn et al., 1984).

NOESY spectra recorded in H$_2$O with a semiselective observation pulse are intrinsically asymmetric, and in addition to NOE's they may manifest exchange of labile protons with the solvent (Haasnoot et al., 1983a). The use of this technique is illustrated in Figure 13.12 for the tetradecamer duplex d(GGAATTGTGAGCGG) · d(CCGCTCACAATTCC) (Boelens et al., 1985b). In this species the base pairs were numbered −2, −1, 1, 2, . . . , 12. With the experimental conditions used the imino-proton resonances −2, −1, 11, and 12 could not be observed. The imino proton–imino proton connectivities for the base pairs 4–9 can be followed in Figure 13.12*A*, where the corresponding cross peaks are well resolved. The only remaining G≡C base pair must be in position 10, but the chemical shift difference relative to G≡C9 is so small that the NOESY cross peak could not be resolved. Similarly, the imino-proton chemical shifts for A═T4 and for the three A═T base pairs that must correspond to positions 1 to 3 are nearly identical, so that the cross peaks could not be identified. Figure 13.12*B* shows the

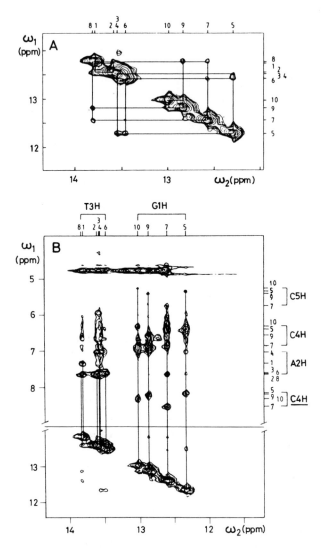

Figure 13.12. Sequential connectivities with labile protons in a *lac*-operator fragment, d(GGAATTGTGAGCGG) · d(CCGCTCACAATTCC) by NOESY (0.008 *M*, H₂O, 0.2 *M* NaCl, 0.02% NaN₃, 0.05 *M* phosphate, pH 6.5, 32°C; 360 MHz; selective observation pulse 45°–τ–45° with τ = 200 μs; sum of four spectra recorded with τ_m = 30, 50, 90, and 150 ms; special base line correction). *(A)* Sequential connectivities between imino protons. *(B)* Connectivities between the imino protons and aromatic or amino protons, where the hydrogen-bonded amino protons of C are underlined. The base pairs are numbered −2, −1, 1, 2, . . ., 12. (from Boelens et al., 1985b).

TABLE 13.2. Chemical Shifts of Labile and Nonlabile Protons in the Watson–Crick Base Pairs 7–11 of the Duplex d(GGAATTGTGAGCGG) · d(CCGCTCACAATTCC)[a]

| Base Pair | Chemical Shifts (32°C, pH 6.5) | | | | |
	G1H or T3H	C4NH$_2^a$	G2NH$_2$	C5H	A2H
GC7	12.31	6.35,8.14	6.38	5.30	
TA8	13.47				7.58
GC9	12.58	6.80,8.51	6.35	5.71	
AT10	13.83				7.61
GC11	12.85	6.53,8.16	6.85	5.36	

[a] In parts per million. The chemical shifts of the hydrogen-bonded amino protons of C are underlined (from Boelens et al., 1985b).

NOESY connectivities from the imino protons to A2H and to the amino protons. The resonance assignments (Boelens et al., 1985a) are indicated on the right, where the hydrogen-bonded amino protons of C are underlined. The band of peaks at $\omega_1 \approx 4.8$ ppm was attributed to exchange of imino and amino protons with H_2O. As an example of the results that can be obtained with the procedures described in this section, Table 13.2 presents the chemical shifts of the protons in the base pairs of part of the duplex studied in Figure 13.12. Combined with the results on the nonlabile protons (Table 13.1), complete assignments for all protons in a DNA duplex can thus be obtained, with the possible exceptions of 5'CH_2, and possibly 4'H and some of the amino protons. As a final comment it may be added that in most DNA duplexes studied, the imino proton–imino proton assignment pathways were interrupted at one or several points by resonance overlap (Figs. 13.11 and 13.12). The assignments were then completed using additional information, such as connectivities in the amino-proton and A2H regions (Fig. 13.9), or observation of different temperature variations of chemical shifts (Fig. 13.1) or of resonance intensities for labile protons in chain terminal and central base pairs during melting of the duplex (Boelens et al., 1985b; Chou et al., 1983; Ulrich et al., 1983; Weiss et al., 1984).

13.4. RESONANCE ASSIGNMENTS IN A-DNA, Z-DNA, AND RNA FRAGMENTS

This section investigates to what extent the assignment procedures for B-DNA can be extended to other nucleic acid conformations. A first general statement on this subject is that those parts of the resonance assignments that rely on scalar coupling connectivities (Chapter 12) can be used with all

different forms of DNA, and with RNA in as far as the ribose-proton resonances can be resolved. In contrast, the situation for sequential assignments
using ^{1}H–^{1}H NOE's must be evaluated separately for each of the different
nucleic acid conformations.

Comparison of A-DNA and B-DNA implies that the same assignment
strategy should work for both conformations. For both forms there are two
short intranucleotide ribose proton–base proton distances. $d_i(6,8;1')$, is
nearly identical in the two forms (Fig. 11.5), and in addition $d_i(6,8;2')$ in B-
DNA and $d_i(6,8;3')$ in A-DNA adopt short values (Fig. 11.6A and C). For
both conformations, short sequential ribose proton-base proton distances
prevail only in the direction 5′ to 3′ of the polynucleotide strands. $d_s(1';6,8)$
and $d_s(2'';6,8)$ are longer in A-DNA than in B-DNA, but in turn $d_s(2';6,8)$ and
$d_s(3';6,8)$ are very short in A-DNA (Table 11.2). The sequential base proton–
base proton distances are similar in the two forms (Table 11.2). If the
NOESY spectra are recorded with a long mixing time, the manifestation of
the differences between A-DNA and B-DNA will be largely masked by spin
diffusion. For the labile protons in the base pairs, the situation is much the
same as for the nonlabile protons. Tables 11.3 and 11.4 imply that the assignment procedures for B-DNA (Section 13.3) are also applicable for A-DNA.

Inspection of the ^{1}H–^{1}H distances in the Z-form of d(CGCGCG)$_2$ provides
some initial indications for the assignment procedures to be used with Z-
DNA. Figures 11.4–11.7 and Tables 11.2–11.4 reveal largely different situations for the nucleotides C and G, and the dinucleotide segments CG and
GC. For C, the intranucleotide distances $d_i(6;1')$ and $d_i(6;2')$ are similar to
the corresponding distances in B-DNA (Figs. 11.5 and 11.6C), whereas for
G, $d_i(8;1')$ is outstandingly short (Figs. 11.5 and 13.3) and $d_i(8;2')$ is longer
than in B-DNA (Fig. 11.6A). In the segment GC there are three short sequential contacts, $d_s(2';5)$, $d_s(2';6)$, and $d_s(3';6)$ (Table 11.2), while there is
no short sequential connectivity in the dinucleotide segment CG. Short sequential distances between nonlabile base protons are absent in both GC and
CG. In as far as the labile protons are concerned, there is a short interstrand
imino proton–imino proton distance in GC, but not in CG. Overall, the
distance data for the Z form of d(CGCGCG)$_2$ thus show that intranucleotide
ribose proton–base proton connectivities in both C and G, and sequential
connectivities in the GC segments are assured, but neither of the assignment
networks used for B-DNA could provide connectivities between subsequent
GC dinucleotide repeats. There are, however, distances linking labile with
nonlabile protons that are sufficiently short for NOE observation, in particular $d_{ps}(C1+,6;G2+,1)$ and $d_{ps}(C1+,1';G2+,1)$ (Table 11.3), which could establish the missing links across the CG junctions. NOE measurements in
H$_2$O solution could thus provide a continuous network of short ^{1}H–^{1}H
distances including backbone protons as well as labile protons in the base
pairs. Compared to B-DNA, a different strategy is thus called for, since
sequential assignments for Z-DNA will probably have to rely on combined
use of data measured in D$_2$O and in H$_2$O.

In practice, resonance assignments for Z-DNA have been obtained by an alternative route, which makes use of the fact that the transition between the B and Z forms of an oligonucleotide duplex can be slow on the NMR time scale. Sequence-specific assignments obtained for the B-DNA form can thus be transferred to the Z form by saturation transfer, which is most conveniently observed by NOESY (Feigon et al., 1984b).

For RNA fragments the same general comments apply as for A-DNA. Limitations on the extent to which assignments of the nonlabile protons can be obtained may arise from the small dispersion of the ribose-proton chemical shifts (Table 2.6). It appears that for short, single-stranded oligoribonucleotides nearly complete ^{1}H NMR assignments can be obtained (Lankhorst et al., 1985). In longer, partly or entirely double-helical RNA fragments, the assignments of nonlabile protons may be limited to the base protons and 1'H of the ribose (Clore et al., 1984; Westerink et al., 1984).

13.5. RESONANCE ASSIGNMENTS IN TRANSFER RNA

Transfer RNA molecules consist of a polynucleotide chain of approximately 80 nucleotides and have a molecular weight of approximately 28,000. They are thus considerably larger than the DNA fragments or the proteins for which complete resonance assignments have been obtained until now. The central region of tRNA ^{1}H NMR spectra is therefore very crowded with resonance lines, and sequential assignments so far concentrated on the hydrogen-bonded imino-proton lines at the low-field end and the methyl resonances in the high-field region of the ^{1}H NMR spectrum (Fig. 3.10). Overall, the procedures used correspond to those for the assignment of the labile protons in DNA fragments (Section 13.3), with certain adaptations for the special situation presented by the structure of tRNA's.

In practice, the atom coordinates in the single-crystal structures of tRNA's were used as the basis for resonance assignments by NOE's (e.g., Heerschap et al., 1982, 1983a,b). The distances between protons in the base pairs of the double-helical stem regions of tRNA are similar to those for A-DNA (Tables 11.3 and 11.4), so that sequential assignments can be obtained along similar lines as in Figures 13.10 to 13.12. Considering the large molecular size, the identification of specified sequence locations as starting positions for the sequential assignments is of decisive importance. This part of the assignment procedure is facilitated by unique local features in tRNA structures. These include approximately 10 nonstandard base pairs in the tertiary structure, and the presence of approximately 10% *rare* nucleotides in the sequence (see Fig. 3.11 for some examples). The latter are responsible for the appearance of well separated methyl lines at the high-field end of the spectrum (Fig. 3.12). The initial NOE experiments with tRNA (e.g., Fig. 6.2) concentrated on the identification of such reference sequence locations (Johnston and Redfield, 1978; Roy et al., 1982a,b; Schejter et al., 1982;

Tropp and Redfield, 1981), in particular by observation of methyl proton–imino proton NOE's.

Similar to the investigations with DNA fragments, assignments with imino proton–imino proton NOE's between adjacent stacked base pairs in tRNA were so far obtained predominantly with TOE difference spectra (Hare and Reid, 1982a,b; Heerschap et al., 1982; 1983a,b; Hyde and Reid, 1985). The example in Figure 6.2 shows that these difference spectra also contain the A2H and amino-proton peaks needed for identification of the base pair type. More recently, 2D experiments were also employed (Hare et al., 1985; Heerschap et al., 1985; Hilbers et al., 1983), and for the future it can be expected that NOESY (Fig. 5.38) will be the preferred technique. Figure 13.13 shows an expanded plot of a small region from a NOESY spectrum of tRNA^Phe in H₂O (Fig. 5.38), which contains cross peaks between the methyl groups of two methylated guanines and hydrogen-bonded imino protons in base pairs. Methylated guanines (Fig. 3.11) are rare bases in tRNA's, so that data of the type of Figure 13.13 can yield sequence-specific assignments for distinct imino-proton resonances.

Similar to tRNA's, sequence-specific assignments for protons in the base pairs can be obtained for other relatively large RNA species. Examples include fragments of 5S and 16S ribosomal RNA from *E. coli* (Heus et al., 1983; Kime and Moore, 1983a).

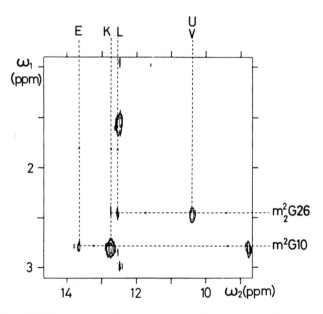

Figure 13.13. NOESY connectivities between methyl groups and amide protons in tRNA^Phe. The region ($\omega_1 - 0.9–3.1$ ppm, $\omega_2 - 8.7–14.6$ ppm) from the spectrum of Figure 5.38 is shown. On the right, sequence-specific assignments are indicated for two methyl resonances. The letters at the top indicate an arbitrary labeling of the imino-proton resonances (from Hilbers et al., 1983).

13.6 NMR DETERMINATION OF THREE-DIMENSIONAL NUCLEIC ACID STRUCTURES

Similar techniques to those described in Chapter 10 for proteins can be applied for the structural interpretation of NMR data on nucleic acids. As with proteins, studies of the global spatial structure and of local conformational properties are best treated in two subsequent, separate phases of a structure determination, which may involve different analyses of the NMR spectra as well as different procedures for structure determination.

Present knowledge indicates that determination of global nucleic acid conformations amounts to identifying one out of a small number of standard structures (Saenger, 1984). The experiments discussed in Section 13.1 can distinguish between flexible, "unstructured" polynucleotides, single-stranded left- or right-handed helical structures with stacked bases, right-handed double helices of the A and B families, and left-handed double helices of the Z family. Using the sequence-specific resonance assignments, a more precise characterization of the helical structures can be obtained even on the basis of a *qualitative* analysis of the ^{1}H NMR spectra. In the following we first consider the experimental data that can be collected as input for a determination of the global conformation.

NOE's with the labile protons in the base pairs have a dominant role in that they manifest long-range or interchain short distances, and provide direct evidence for the hydrogen bonds present in the molecular structure. The situation is similar to that with β sheets in proteins (Chapter 9): Once sequence-specific resonance assignments are available, the two polynucleotide segments involved in the formation of a double-helical structure can be identified from these long-range NOE's. From the patterns of NMR lines corresponding to the hydrogen-bonded imino protons (e.g., Fig. 3.8) further inferences can then be made on the *symmetry* of the double-helical segments.

In the individual polynucleotide strands, evidence for distinguishing between different conformations can be obtained from qualitative interpretation of sequential and intranucleotide NOE's between nonlabile protons (i.e., differentiation between *strong, medium,* and *weak* NOE's in the same way as described for proteins in Table 10.1). The NOE corresponding to the outstandingly short distance $d_i(8;1')$ for purine nucleotides adopting the syn form in Z conformations was already discussed (Fig. 13.3). The distances $d_i(6,8;2')$ and $d_s(2'';6,8)$ are very short in B-DNA, and $d_i(2'';4')$, $d_i(6,8;3')$, $d_s(2';6,8)$, and $d_s(3';6,8)$ are very short in A-DNA (Figs. 11.4 and 11.6, Table 11.2). Measurement of the relative intensities of the NOESY cross peaks corresponding to these distances should therefore allow us to distinguish between A- and B-DNA, provided that the NOESY spectra used for this purpose are recorded with very short mixing times (Figs. 13.5 and 13.6). The small number of nucleic acid single-crystal structures determined at high resolution does not presently warrant a statistical analysis for estimating the extent and uniqueness of the identification of A- and B-DNA by such qualitative distance criteria (see Section 9.1). However, inspection of the available crystal structures indicates that the changes in the ^{1}H–^{1}H distances corresponding to the sequence-dependent variations of the helix parameters observed in single crystals (Calladine, 1982; Dickerson, 1983) do not result in overlap of the ranges for $d_i(2'';4')$, $d_i(6,8;2')$, $d_i(6,8;3')$, $d_s(2';6,8)$, $d_s(2'';6,8)$, and $d_s(3';6,8)$ in the A-DNA and B-DNA families. The distinction between these two conformations on the basis of semiquantitative NOESY data should thus be quite unambiguous. Further evidence for identifying A- and B-DNA may be obtained from the characteristic differences between certain vicinal ^{1}H–^{1}H coupling constants (Altona, 1982), which manifest the different dominant ring puckers in the two conformations (Table 11.5). For the 3'C-endo deoxyribose conformation in A-DNA, J values are of the order $^3J_{1'2'} = 2$ Hz, $^3J_{1'2''} = 8$ Hz, $^3J_{2'3'} = 7$ Hz, $^3J_{2''3'} = 9$ Hz, and $^3J_{3'4'} = 8$ Hz, and for the 2'C-*endo* form in B-DNA the corresponding values $^3J_{1'2'} = 10$ Hz, $^3J_{1'2''} = 6$ Hz, $^3J_{2'3'} = 6$ Hz, $^3J_{2''3'} = 1$ Hz, and $^3J_{3'4'} = 1$ Hz. In practice, semiquantitative assessment of the couplings in the fragment 1'CH–2'CH$_2$–3'CH of the deoxyribose rings, in particular $^3J_{1'2'}$, $^3J_{2''3'}$, and $^3J_{3'4'}$ (Fig. 12.3) should provide a reliable criterion for distinguishing between A- and B-DNA. Finally, the experimental chemical shifts may be checked for compatibility with the predictions from ring-current calculations using the atom coordinates for standard A- or B-DNA (or other conformations) (Bell et al., 1985; Giessner-Prettre, 1984).

The information on upper distance constraints between distinct hydrogen atoms, on the hydrogen bonds, and on spin–spin coupling constants could be used for determination of the global polynucleotide conformation either by an empirical model-building approach (Section 10.2) or by distance geometry calculations (Section 10.3). So far, model building using mechanical molecular models or computer graphics has been used. In many instances, however, the structure determinations were limited to finding a data set compatible with one of the standard DNA or RNA conformations (e.g., Fig. 11.8), so that no new modeling of the structure was called for. Examples of identification of B-type double helices are d(GGTATACC)$_2$ (Reid et al., 1983), d(ATATCGATAT)$_2$ (Feigon et al., 1982), d(CGCGAATTCGCG)$_2$ (Hare et al., 1983), d(GGAATTGTGAGCGG) · d(CCGCTCACAATTCC) (Boelens et al., 1985) and d(AAAGTGTGACGCCGT) · d(ACGGCGTCACACTTT) (Gronenborn et al., 1984). A single-strand B-type helix was proposed for d(AAGTGTGATAT) (Clore and Gronenborn, 1984), and an A-type double helix was found for r(CGCGCG)$_2$ (Westerink et al., 1984). r(CGCGCG)$_2$ was also compared with d(CGCGCG)$_2$ at low salt concentration (see Fig. 13.2), and the two compounds were found to adopt A- and B-type conformations, respectively (Haasnoot et al., 1984).

Refinements beyond identification of the structure type are obviously of interest, for example, to investigate if the sequence-dependent structure variations predicted by Calladine, 1982 also prevail in the solution conformations of nucleic acids. Initial evidence for sequence-dependent base-pair stacking in a DNA fragment in solution was obtained from studies of steady-state NOE's between adenine 2H in neighboring base pairs of d(CGATTA-TAATCG)$_2$ (Patel et al., 1983b). This experiment made use of the facts that in D$_2$O solution the 2-protons of adenine are isolated from close contact with other protons (Fig. 11.1), and that the interbasepair, interstrand A2H–A2H distances can be markedly affected by propeller twisting of the base pairs (Fig. 11.9). For more comprehensive refinements including the polynucleotide-backbone conformations, initial applications of the following two approaches have been described.

To investigate correlations between specified DNA conformations and the corresponding NOESY spectra, complete relaxation matrices taking account of all proton–proton dipolar couplings and allowing for spin diffusion were considered (Keepers and James, 1984). Using this formalism, NOESY spectra at different mixing times were computed for d(GGAATTCC)$_2$ (Broido et al., 1985) and d(GGTATACC)$_2$ (Jamin et al., 1985), assuming, for example, either a standard B conformation or an energy-refined B conformation. These theoretical spectra were then compared with NOESY experiments recorded with variable, relatively long mixing times. From observation of similarities as well as differences between the experiments and the theoretical spectra, sequence-dependent variations of structure and effective correlation times were inferred. Eventually, such formalisms should ideally be used as a *final check* for internal consistency of structures determined from the NMR data, rather than as a trial-and-error method for structure determination. They might also be valuable for systematic studies of the manifestations of specified structural features in the NOESY spectra.

A second refinement technique used with DNA and RNA fragments is related to the refinement procedure applied by Kaptein et al., 1985 with the *lac*-repressor DNA-binding domain (Section 10.5). A trial conformation (e.g., standard B-DNA) was subjected to a restrained least-squares refinement. Thereby, the NOE experiments were interpreted in terms of relatively accurate ^{1}H–^{1}H distances (e.g., $\lesssim \pm 0.2$ Å in Clore et al., 1985b), which are used as the constraints in the refinement procedure. For two DNA-duplexes studied, d(CGTACG)$_2$ (Clore et al., 1985b) and d(AAGTGTGACAT) · d(ATGTCACACTT) (Clore and Gronenborn, 1985a), B-DNA conformations with local structure variations were reported. For a RNA duplex, r(CACAG) · r(CUGUG) (Clore et al., 1985a), an A-type helix with local structure variations was reported after application of the same treatment.

PART IV
WITH NMR TO BIOPOLYMER CONFORMATION AND BEYOND

CHAPTER **14**
Conformation of Noncrystalline Proteins and Nucleic Acids

This chapter mainly involves some comments on the significance of the molecular conformations that can be determined with NMR in solution. Closely related to this main theme are brief discussions on the selection of proteins and nucleic acids for which high-resolution NMR studies are feasible, and on the scope of NMR investigations for different classes of compounds. In addition, Section 14.2 considers some practical aspects of structure determinations by NMR.

14.1. PROJECTS FOR STRUCTURE DETERMINATION BY NMR

Detailed NMR studies are feasible for proteins and nucleic acids with well resolved NMR spectra containing a large number of separated, sharp resonance lines. Primarily, this requirement imposes bounds on the size, which are, however, not rigid limitations. Furthermore, considering that 2D NMR spectra are recorded in relatively concentrated solutions, the compounds must be available in milligram quantities, and good solubility without self-aggregation at high concentration is a quite stringent selection criterion. The compounds should also be stable in solution over periods of several days.

Nearly all illustrations in this book show experiments with compounds in the molecular weight range 5,000-15,000. [In many instances the results of the NMR studies relate, however, to the biological functions of larger compounds. Examples are the structure determinations with synthetic DNA fragments, and with protein domains isolated by selective cleavage of multidomain structures (e.g., Zuiderweg et al., 1984b; De Marco et al., 1985).] On a limited scope, structural investigations by NMR in solution can also be conducted with larger molecules. Examples include a variety of proteins, the tRNA's (Figs. 3.10 and 3.12), polyribonucleotides (Broido and Kearns, 1982), and polydeoxyribonucleotides (Borah et al., 1985). (Using different principles, very large particles can be investigated with solid-state NMR methods; e.g., Cross and Opella, 1985.)

The intrinsic properties of the molecules studied can largely influence the scope of a structure determination by NMR and the amount of work involved. In the following this is exemplified with some representative situations.

In an investigation of a new protein, one is first of all faced with the task of characterizing a new type of conformation. This requires collection of the maximum possible number of conformational constraints to define the global features of the secondary and tertiary structure. The work load for the resonance assignments and the determination of the global 3D structure with proteins of molecular weight 5,000-10,000 has so far been of the order of one to several man years. Considerably improved efficiency should in the future be achieved with partial automation of the spectral analysis.

NMR is a particularly powerful method for comparative studies of groups of related molecular structures. Therefore, once the NMR spectra of a particular protein have been analyzed, the investigations may be efficiently extended to homologous proteins. Following the results previously obtained for the parent compound, sequence-specific resonance assignments and identification of the regular secondary structure may then typically be obtained with a few day's to a few week's work (e.g., Chazin et al., 1985; Stassinopoulou et al., 1984; Štrop et al., 1983b). This offers interesting possibilities for systematic studies of correlations between covalent structure and conformation, or also for practical applications in protein product control. It should be quite straightforward to develop automated procedures for such structure comparisons.

Studies of nucleic acid fragments generally benefit from the outstanding ability of NMR to conduct comparative investigations of conformational analogs, since in most instances one faces the task of identifying one out of a small selection of conformation types. Compared to proteins, determination of the global conformation is usually a less important part of the project, and serves primarily as a starting point for structure refinement, studies of the molecular dynamics (Section 14.3), or investigations of intermolecular interactions (Chapter 15). Combined with the nearly unlimited possibilities for designing new structures by chemical synthesis, NMR is a powerful and efficient method for large-scale investigations of the influence of covalent-structure variations in nucleic acids on conformational and functional properties (e.g., Haasnoot et al., 1983b; Lankhorst et al., 1985; Patel et al., 1985a,b).

By variation of the solution conditions, the scope of NMR investigations can be extended beyond studies of the native conformations in aqueous solution. This may include, for example, biopolymers bound to lipid or detergent micelles (Arseniev et al., 1985; Braun et al., 1981, 1983; Brown et al., 1981, 1982), solutions in nonaqueous media, or folding intermediates in denaturation studies, which may be trapped under certain conditions of solvent and temperature (e.g., Haasnoot et al., 1983c; States et al., 1984). For such projects, the potential of NMR for providing precise structural information is unique among presently available experimental techniques.

14.2. PLANNING A STRUCTURE DETERMINATION BY NMR

High-field NMR equipment (presently 360–500 MHz for ¹H) is indispensable for ¹H NMR studies of proteins and nucleic acids. Otherwise, all experiments described in this book can in principle be recorded with commercially available instrumentation, which is usually also equipped with sufficient data-storage capacity. It is helpful to have extra computer facilities available for data manipulation and spectral analysis. In our experience, for each structure determination the spectrometer should be available for one or several uninterrupted periods of 5 to 10 days to record the 2D NMR spectra (see paragraph *2*). For the structural interpretation of the data, access to ample computing facilities and a computer-graphics system should be arranged.

Once a particular molecule has been selected, three phases can be distinguished in the actual structure determination:

1. The first phase consists of an extensive search for optimal solution conditions for the NMR experiments. For proteins this typically includes studies of the stability of the native conformation in H_2O over a wide range of pH, temperature, and ionic strength and determination of the mobility of the aromatic rings and the protonation–deprotonation equilibria for distinct ionizable groups. (For example, if the stability permits, one may want to work at an elevated temperature because at higher temperatures the resonance lines are usually narrower. One must also consider, however, that dynamic processes may lead to line-broadening at high temperature (Fig. 1.5). Another viewpoint may be to select conditions resembling the physiological environment of the protein.) With nucleic acids one may want to define solution conditions where a specified form prevails, for example, single strands or a duplex (Fig. 13.1). For this first phase, 1D NMR (Figs. 1.5, 3.3, 3.6–3.8, 3.10, 3.12, 3.13) is most efficient, since a large number of measurements with different conditions must be recorded. Inspection of the 1D ¹H NMR spectra may also give initial indications on the prevalent conformations, which can be useful for the subsequent work. [For example, in a DNA duplex the number of imino-proton lines can reflect conformational symmetry (Fig. 3.8). In proteins, slow amide-proton exchange is characteristic for the hydrogen bonds in regular secondary structures (Fig. 3.3*A*), and low field shifts of backbone NH and αH resonances are indicative of β sheets (Pardi et al., 1983b; Wagner et al., 1983)]. For proteins, this exploratory phase should include examination of COSY fingerprints in H₂O (Figs. 5.9B, 5.11, 8.12, 8.19) to further assess the homogeneity of the protein preparation, and the spectral resolution under the solution conditions selected for the structure determination.

2. In a second phase, the 2D NMR spectra are recorded. Ideally, two groups of 4 to 10 experiments (depending on the complexity of the structure) recorded with identical conditions in D₂O and H₂O solution should provide all the data needed for the resonance assignments and the structure determination. These spectra must be directly comparable, which is best achieved by recording the experiments in uninterrupted series, without removing the sample from the spectrometer. For automatic correction of slow instrumental drifts, all spectra might be recorded in parallel, in an *interleaved mode*.

3. The third phase includes the spectral analysis, resulting in the sequence-specific resonance assignments (Figs. 8.18 and 8.19; Tables 8.3, 13.1, 13.2) and the input for the determination of the 3D structure, which consists of NOE distance constraints, spin–spin coupling constants, identification of hydrogen bonds, and possibly other parameters (Figs. 9.3–9.6, 10.1, 10.2; Table 10.3). Also included is the structural interpretation of these data, using the methods described in Chapters 9 and 10 and Section 13.6. In practice, additional NMR experiments are often planned on the basis of the ongoing spectral and structural interpretations. To facilitate direct comparison with other measurements, some or all of the previously recorded experiments may then in some instances be repeated, which brings about additional demands for spectrometer time.

14.3. STRUCTURE AND DYNAMICS OF PROTEINS AND NUCLEIC ACIDS IN SOLUTION

Long before sequence-specific assignments became available and one could tackle 3D structure determinations of proteins or nucleic acids by NMR, a variety of NMR experiments provided insight into intramolecular rate processes in these macromolecules. [For example, observation of segmental chain mobility in the spin-relaxation parameters, measurements of aromatic ring flips (Figs. 1.5, 1.6, and 3.4), and amide-proton exchange studies (Figs. 3.13, 3.14, 5.35); for recent reviews see Jardetzky and Roberts, 1981; Kearns, 1984; Wagner, 1983b; Wüthrich and Wagner, 1984.] Combined with sequence-specific resonance assignments, maps of parameters manifesting structure fluctuations on different time scales can thus be obtained, which extend over entire protein molecules (e.g., Fig. 1.3; Boelens et al., 1985; Wüthrich et al., 1984b) and nucleic acid fragments (e.g., Chou et al., 1984; Figueroa et al., 1983; Lee et al., 1985; Patel et al., 1983a). Hydrogen-exchange studies, which have long assumed a dominant role among the experimental techniques for investigations of internal mobility in proteins and nucleic acids, were revolutionized by the ability of NMR to provide quantitative exchange rates for distinct, individual protons (Figs. 3.13 and 5.35; Englander and Kallenbach, 1984; Roder et al., 1985; Tüchsen and Woodward, 1985b; Wagner, 1983b; Wagner et al., 1984; Woodward et al., 1982). The extensive experimental documentation of internal motility of proteins and nucleic acids in the solutions used for NMR experiments provides the background for the following reflections on the significance of molecular conformations determined by NMR.

For both polypeptides and polynucleotides one may encounter different limiting situations, which require different treatment of NMR data relating to the molecular conformation. In one of these, the chain segments are locked in unique spatial arrangements, with intramolecular motions restricted to thermal fluctuations about the coordinates corresponding to this conformation. Alternatively, there may be equilibrium situations with comparable populations for two or several conformers. The rate of exchange between the different conformers in such a system can vary over a wide range and may be slow or rapid on the NMR time scale, so that one observes either separate resonances for the individual species, or a single, average NMR spectrum.

An equilibrium situation with multiple conformers is typically encountered for short linear peptides and nucleotides in solution. A direct visualization of these equilibria is obtained for proline-containing peptides, since the exchange between species containing *cis*- and *trans*-proline is sufficiently slow to enable observation of two different NMR spectra for each pair of isomers (e.g., Wüthrich, 1976). Usually, however, a single spectrum is observed, which corresponds to the average of the spectra of the individual conformers. The NMR parameters measured in this situation correspond to an average of the parameters in the individual conformers weighted by the populations of these species. For a large number of oligopeptides and oligonucleotides, the results of conformational studies based on analyses of chemical shifts and spin–spin coupling constants were presented in the form of a population analysis for distinct rotamers about single bonds. (A recent example is Mellema et al., 1984; for reviews see, e.g., Altona, 1982; Jardetzky and Roberts, 1981.) In such systems a structural analysis of NOE's could be misleading, since admixture of a small population of conformers with outstandingly short distances for distinct pairs of protons could have a dominant effect on the observed, average spectrum. However, the ^{1}H–^{1}H NOE's are usually quenched by the pronounced flexibility of such molecules, so that the potentially misleading NOE information is in practice not accessible with the commonly used experiments. Similar situations to that outlined here for small peptides and nucleotides may also prevail for larger molecules. For example, polypeptide hormones tend to adopt nonglobular, flexible conformations in solution. Distinct, well-defined conformations of such molecules may then be trapped in aggregates with other species (in the physiological environment, this may be a membrane surface or a specific receptor). This is exemplified by glucagon, a polypeptide of 29 residues, which adopts a predominantly flexible, monomeric form in dilute aqueous solution (Bösch et al., 1978), and a well-defined conformation when bound to the surface of lipid micelles (Figs. 5.11 and 10.6; Braun et al., 1981, 1983).

The available evidence shows that the interior parts of globular proteins and double-helical nucleic acids in solution are usually locked in unique spatial arrangements. Theoretical investigations of the influence of high-frequency motions in globular proteins on the ^{1}H–^{1}H NOE's and on the vicinal spin–spin coupling constants were recently reported (Hoch et al., 1985; Olejniczak et al., 1984). They support the assertion that NOE determination of distance bounds with the assumption of structural rigidity, or using simple uniform averaging to account for the molecular dynamics (Fig. 6.14), is a justifiable approach for determination of the global spatial structure in these molecular regions (Chapter 10). More comprehensive treatments of the problems associated with NOE distance measurements in dynamic molecules, for example, from evaluation of complete relaxation matrices (Keepers and James, 1984) may in the near future provide a basis for further quantitation of distance measurements by NOE's, which will be needed for structure *refinements*.

For globular proteins, the ^{1}H–^{1}H NOE's also indicate qualitatively different situations for amino acid residues in interior regions and on the molecular surface. In contrast to the entire polypeptide backbone (with the possible exception of the chain ends) and the amino acid side chains in the interior of the molecule, which give negative ^{1}H–^{1}H NOE's and are accessible for the structure-determination procedures of Chapters 9 and 10, the surface side-chain protons beyond βCH_2 are often NOE invisible. Alternative procedures must therefore be applied to further characterize the molecular surface. Initially one may want to distinguish between surface areas where the amino acid side chains represent *dangling ends* characterized by comparable populations of different staggered rotamers about the single bonds, and regions where the side chains are locked in unique spatial arrangements by interactions with backbone segments or with other side chains. Presently available experiments for this purpose include analysis of the vicinal ^{1}H–^{1}H coupling constants in the $\alpha CH–\beta CH_2$ fragments (Hoch et al., 1985; Nagayama and Wüthrich, 1981b), measurement of the pH dependence of amide-proton chemical shifts (Bundi and Wüthrich, 1979b; Ebina and Wüthrich, 1984), and a variety of "classical" approaches including analyses of chemical shifts and manifestations of close approach between charged groups in the protonation–deprotonation equilibria (e.g., Dwek, 1973; Jardetzky and Roberts, 1981; Wüthrich, 1976). Quite possibly, the recently proposed *rotating frame* NOE experiment (Bothner-By et al., 1984) might be useful for studies of flexible amino acid side chains on the protein surface (Marion, 1985). Overall, however, the development of a methodology for comprehensive characterization of the protein surface in solution is left as an exciting challenge for the future.

The following are some consequences of these considerations on the conformational states in globular proteins:

1. In general, the two limiting principles of polypeptide segments either locked in unique spatial arrangements or adopting equilibrium situations with multiple populated conformers are both represented in the same molecule. Qualification of a structure determination in solution may therefore include specification of the accuracy of the atom positions in the molecular core, as well as a statistical description of the surface.

2. When comparing corresponding structures determined in solution and in crystals, one can in many instances anticipate close coincidence for the core regions (e.g., Fig. 10.10). In contrast, surface side chains which are flexible in the solution conformation may be locked in unique orientations in the crystals, so that in the crystal structure the positioning of such segments is more precise (but not necessarily more relevant for studies of structure–function correlations).

3. In view of the dominant role of the surface residues for most aspects of biological functions, detailed studies of the protein surface in solution should have high priority in future endeavors. Systematic studies of small proteins could provide information on local surface conformations formed by distinct combinations of amino acid residues, which might then be applied (in conjunction with the crystal structure) for characterization of the surface in proteins that are too big for detailed direct investigation by NMR in solution.

In double-helical nucleic acid fragments, negative NOE's are commonly observed for the bases in the interior as well as the polynucleotide backbone on the surface. A priori there is no reason to assume the presence of highly mobile structural segments, except possibly those due to fraying at the chain ends. Nonetheless, evidence for conformational equilibria between multiple, significantly populated states, has been implicated from different lines of evidence. For example, it was concluded from ^{31}P NMR studies that only the central nucleotide pairs in double-helical oligonucleotide fragments adopt conformational rigidity comparable to polynucleotide duplexes (Connolly and Eckstein, 1984). For d(CGCGCG)$_2$ in the B form, evidence was presented for conformational heterogeneity in the pucker of the deoxyribose rings, with significant populations of both N and S states (Fig. 11.3), and it was suggested that conformational equilibria between different ring puckers are a general phenomenon in intact B-DNA duplexes (e.g., Haasnoot et al., 1984). These observations could well be a starting point for further investigations on the significance of structure refinements relying on quantitative NOE measurements of ^{1}H–^{1}H distances with hydrogen atoms of the ribose rings.

CHAPTER 15
NMR Studies of Intermolecular
Interactions with Biopolymers

Studies of intermolecular interactions had a prominent role among the early biological applications of NMR, and with the availability of the NMR experiments described in the preceding chapters, a renewed upsurge of activities in this area is observed. The range of potential applications is almost boundless and includes topics of prime interest, such as DNA-drug interactions in cancer chemotherapy and regulation of nucleic acid functions by specific interactions with proteins. Using strategies along the line indicated by Figure 1.2, modern NMR techniques can be expected to contribute detailed information on structural, thermodynamic, and kinetic aspects of the intermolecular interactions in such systems.

From the viewpoint of NMR spectroscopy, different procedures may be called for, depending on the thermodynamic and kinetic stability of the intermolecular aggregates. We first consider the two limiting situations of rapid exchange of the interacting components between free and complexed form, and of stable systems without exchange. This then provides a basis for examination of intermediate situations, for example, with two or several interacting molecules forming moderately stable aggregates, or with a ligand molecule exchanging between multiple binding sites of a macromolecule.

Aqueous solutions of proteins or nucleic acids present suitable examples of low-stability intermolecular aggregates. In molar units they contain typically a 10^4–10^5 fold excess of water relative to the macromolecular solute. At any given moment a small percentage of the water molecules is bound as *hydration water* on the macromolecular surface, and the mobility of the bound water is largely restricted to the Brownian motions of the macromolecules. The effective rotational correlation time for dipole–dipole coupling between the water protons may thus be increased from approximately 2×10^{-12} s to a value near 10^{-8} s, which corresponds to a line broadening by a factor 10^3–10^4 (Fig. 3.1). Because of the low concentration relative to the bulk water, the NMR line of the bound water cannot usually be observed. However, if the exchange of water molecules in and out of the hydration sites is rapid compared to the spin relaxation in the free solvent, the spin states acquired during the complexation with the macromolecule persist after the exchange into the bulk water, which causes a *relaxation enhancement* for the solvent resonance. This relaxation enhancement is governed either by the relaxation rate for the bound water, by the rate of exchange between hydration site and bulk water, or by a combination of the two, depending on the relative rates of the individual processes. The observed relaxation enhancement may thus yield information on the macromolecular surface, for example, the number of hydration sites and the degree of immobilization of the bound water molecules, on the exchange kinetics, or on both. Application of these principles can be extended to other molecules added to the solution in high concentration relative to that of the macromolecules, for example, small organic compounds, or fluoride ions. Two attractive features of such experiments are that the macromolecule may be present in much lower concentration than for direct NMR observation, and that the information is obtained from a simple NMR spectrum consisting of a single line or a small number of resolved resonances. In the 1960s, before NMR equipment for high-resolution studies of biopolymers was available, relaxation-enhancement studies were extensively applied to a variety of biomacromolecular systems, for example, by Cohn, Jardetzky, Koenig, Mildvan, Shulman, and their colleagues. (For reviews see Dwek, 1973; James, 1975; Jardetzky and Roberts, 1981; Wüthrich, 1970, 1976.)

A modern application of the principles of relaxation enhancement in multicomponent systems with exchange, which makes use of the potentialities of modern NMR equipment, are *transferred NOE* experiments (Albrand et al., 1979; Clore and Gronenborn, 1982, 1983). If small molecules containing two or several protons are transferred from the free state in solution to a binding site on a macromolecule, the rotational correlation time effective for intramolecular ^1H–^1H NOE's changes from a regime with positive NOE's to a situation where the intramolecular NOE's have negative sign (Fig. 6.4). Furthermore, the molecules in the bound state may adopt different conformations from those in the free state. By the exchange between the macromolecular binding site and the free state, the NOE's characterizing the conformation of the bound molecule can be transferred to more easily observable resonances, which correspond either to the spectrum of the free molecule modulated by the exchange with the macromolecular binding sites, or to the average of the spectra for the bound and free molecules. Examples include studies on the conformations of nucleoside-triphosphate effectors bound to the quite large proteins hemoglobin and aspartate transcarbamylase (Banerjee et al., 1985; Gronenborn et al., 1984a).

Hemoproteins may serve as an illustration for stable intermolecular aggregates. Their structural entity involves interactions of the polypeptide chains with nonpeptide components, the heme groups (Fig. 10.5). The stability of these aggregates is such that hemoprotein solutions are devoid of detectable concentrations of apoprotein or free heme groups. For heme groups in diamagnetic hemoproteins, individual ^{1}H NMR assignments can be obtained in a sequential fashion, using NOE's between the neighboring groups of protons attached on the periphery of the ring (Fig. 10.5) (Keller and Wüthrich, 1978, 1981). Following the scheme of Figure 1.2, additional ^{1}H–^{1}H NOE's between heme protons and distinct hydrogen atoms of the polypeptide chain can then be used for detailed investigations of the environment of the heme, which usually constitutes the active site in these proteins (e.g., Fig. 10.5; Keller and Wüthrich, 1980, 1981; La Mar, 1979; Moore et al., 1985).

The stability of a wide variety of biologically interesting complexes with proteins and nucleic acids is somewhere between the loose association with the solvent water and the nearly inert aggregates formed by hemoproteins. Typically, the interacting species undergo reversible associations and the stability of the complexes may vary over a wide range, depending on solution conditions such as pH, ionic strength, and temperature. NMR may then first be used to monitor the thermodynamic and kinetic stability of the complexes formed under different experimental conditions, and to define suitable solution conditions for more detailed studies, for example, of the spatial structure of the aggregated particles (an illustration is presented by the thermal melting curves for a DNA duplex in Figure 13.1). Ideally, for complexes with high stability, nearly homogeneous solutions containing only very low concentrations of the free components might thus be prepared. Notwithstanding additional complexity that may arise from the presence of these additional components and from exchange phenomena in the NMR spectra (an illustration of NMR manifestations of dynamic processes with slow and intermediate rates is presented by the temperature-dependent variations in Fig. 1.5), structural studies of such complexes can then in principle be pursued along similar lines as described for free proteins or nucleic acids (Chapters 9 and 10, and Section 13.6).

Some typical questions for NMR experiments are to compare the conformations of the constituent molecules in the free form and in the aggregates, and to identify the regions of intermolecular contacts. Initial indications for the location of these contact sites can be obtained from investigations of the effects of complexation on intrinsic properties of the individual components, such as the chemical shifts (either of the naturally present hydrogen or phosphorous nuclei, or of isotopes such as ^{13}C, ^{15}N, or ^{19}F, which are introduced into the system for the purpose of these NMR experiments), the spin–spin couplings, the spin relaxation times, or other parameters manifesting global or local dynamics of the molecules. In a different approach, the accessibility to the solvent or to other solute molecules of specified molecular segments in the free and complexed polymer chains may be compared. For example, photochemically induced dynamic nuclear polarization (photo CIDNP) probes the accessibility of aromatic rings in the biopolymers to dye molecules, which are added to the solution for this experiment (Buck et al., 1980; Kaptein et al., 1979). As an alternative, the accessibility of surface groups to paramagnetic shift reagents could be studied (Dwek, 1973; Wüthrich, 1976; Jardetzky and Roberts, 1981). The most direct information, however, can again be obtained from NOE experiments. These can be used to probe the conformations of the individual molecules in the complexes, as well as to delineate close proximity of nuclear spins in the different interacting molecules. It should be emphasized again that with all these experiments the information obtained will largely depend on the availability of sequence-specific resonance assignments for the interacting molecules in the complex (Fig. 1.2).

Experiments with complexes formed between DNA duplexes and a variety of drug molecules present nice illustrations for practical applications of these principles. Quite generally it was found that the DNA-duplex structure is preserved in such aggregates. Sequence-specific ^{1}H NMR assignments in the complex then provide spectroscopic probes for distinguishing between different modes of drug binding. For example, Figure 11.1 shows that the minor groove contains only the nonlabile base protons in position 2 of adenine, whereas the major groove is lined by the protons in position 8 of the purine bases and in the positions 5 and 6 of pyrimidines. Qualitative analysis of intermolecular ^{1}H–^{1}H NOE's can therefore provide direct evidence for distinguishing between *intercalating* drugs, which are sandwiched between neighboring stacked base pairs, and drugs binding to the outside of the DNA, lining either the major or minor groove. [Independent evidence for distinguishing between intercalating drugs and outside binding drugs can be obtained from the imino-proton chemical shifts: Intercalation causes upfield shifts, outside binding causes downfield shifts (e.g., Feigon et al., 1984a).] Some examples are studies of the interactions of the intercalating agent actinomycin and the minor groove binding agent netropsin to the DNA duplexes d(ATGCAT)$_2$ and d(CGCGAATTCGCG)$_2$ (Brown et al., 1984; Pardi et al., 1983c; Patel, 1979, 1982; Patel et al., 1982a).

Initial experiments (e.g., Alma et al., 1983; Buck et al., 1983) indicate that experimental approaches similar to those used for studies of DNA–drug complexes should be applicable for studies of protein–nucleic acid interactions. The increased complexity of the NMR spectra in such higher molecular weight aggregates makes detailed studies more difficult, but investigations of protein–nucleic acid interactions are undoubtedly one of the most exciting prospects for NMR applications in the near future. This is clearly indicated by numerous recent publications on NMR studies with DNA duplexes corresponding to protein recognition sites, and with DNA-binding proteins (e.g., Figs. 10.2–10.4; Lane and Jardetzky, 1985; Lefevre et al., 1985; Kaptein et al., 1985; Weber et al., 1985b; Wemmer et al., 1985).

This short chapter can hardly do justice to the outstanding potentialities of NMR experiments for studies of intermolecular interactions with biopolymers. In particular with nucleic acids, these might well turn out to be the biologically most relevant use of sequence-specific resonance assignments. It is a big advantage that NMR allows extensive serial studies with variation of the solution conditions, or with systematic modifications in the chemical structures of the interacting molecules. This should allow to expand the scope of such projects beyond the experimental limits for techniques that depend on the use of single crystals for studies of intermolecular interactions.

References

Abragam, A. (1961). *The Principles of Nuclear Magnetism*, Clarendon, Oxford.

Albrand, P. L., B. Birdsall, J. Feeney, G. C. K. Roberts, and A. S. V. Burgen (1979). *Int. J. Biol. Macromol.*, **1**, 37–41.

Alma, N. C. M., B. J. M. Harmsen, J. H. van Boom, G. van der Marel, and C. W. Hilbers (1983). *Biochemistry*, **22**, 2104–2115.

Altona, C. (1982). *Recl. Trav. Chim. Pays-Bas*, **101**, 413–433.

Altona, C. and M. Sundaralingam (1972). *J. Am. Chem. Soc.*, **94**, 8205–8212.

Anil-Kumar, R. R. Ernst, and K. Wüthrich (1980a). *Biochem. Biophys. Res. Comm.*, **95**, 1–6.

Anil-Kumar, G. Wagner, R. R. Ernst, and K. Wüthrich (1980b). *Biochem. Biophys. Res. Comm.*, **96**, 1156–1163.

Anil-Kumar, G. Wagner, R. R. Ernst, and K. Wüthrich (1981). *J. Am. Chem. Soc.*, **103**, 3654–3658.

Anil-Kumar, R. V. Hosur, and K. Chandrasekhar (1984). *J. Magn. Reson.*, **60**, 143–148.

Armitage, I. M. and J. D. Otvos (1982). "Principles and Applications of ^{113}Cd NMR to Biological Systems," in *Biological Magnetic Resonance*, L. J. Berliner and J. Reuben, Eds., Vol. 4, Plenum, New York, pp. 79–144.

Arseniev, A. S., G. Wider, F. J. Joubert, and K. Wüthrich (1982). *J. Mol. Biol.*, **159**, 323–351.

Arseniev, A. S., V. I. Kondakov, V. N. Maiorov, T. M. Volkova, E. V. Grishin, V. F. Bystrov, and Yu. A. Ovchinnikov (1983). *Bioorgan. Khim.*, **9**, 768–793.

Arseniev, A. S., V. F. Bystrov, V. T. Ivanov, and Yu. A. Ovchinnikov (1984a). *FEBS Lett.*, **165**, 51–56.

Arseniev, A. S., V. I. Kondakov, V. N. Maiorov, and V. F. Bystrov (1984b). *FEBS Lett.*, **165**, 57–62.

Arseniev, A. S., I. L. Barsukov, V. F. Bystrov, A. L. Comize, and Y. A. Ovchinnikov (1985). *FEBS Lett.*, **186**, 168–174.

Aue, W. P., E. Bartholdi, and R. R. Ernst (1976a). *J. Chem. Phys.*, **64**, 2229–2246.

Aue, W. P., J. Karhan, and R. R. Ernst (1976b). *J. Chem. Phys.*, **64**, 4226–4227.

Bachmann, P., W. P. Aue, L. Müller, and R. R. Ernst (1977). *J. Magn. Reson.*, **28**, 29–39.

Baillargeon, M. W., M. Laskowski, D. E. Neves, M. A. Porubcan, R. E. Santini, and J. L. Markley (1980). *Biochemistry*, **19**, 5703–5710.

Bain, A. O. (1984). *J. Magn. Reson.*, **56**, 418–427.

Banerjee, A., H. R. Levy, G. C. Levy, and W. W. C. Chan (1985). *Biochemistry*, **24**, 1593–1598.

Bartholdi, E. and R. R. Ernst (1973). *J. Magn. Reson.*, **11**, 9–19.

Basus, V. J. (1984). *J. Magn. Reson.*, **60**, 138–142.

Baumann, R., G. Wider, R. R. Ernst, and K. Wüthrich (1981). *J. Magn. Reson.*, **44**, 402–406.

Bax, A. (1982). *Two-Dimensional Nuclear Magnetic Resonance in Liquids*, Reidel, London.

Bax, A. and D. G. Davis (1985). *J. Magn. Reson.*, **65**, 355–360.

Bax, A and G. Drobny (1985). *J. Magn. Reson.*, **61**, 306–320.

Bax, A. and R. Freeman (1981). *J. Magn. Reson.*, **44**, 542–561.

Bax, A., A. F. Mehlkopf, and J. Smidt (1979). *J. Magn. Reson.*, **35**, 373–377.

Bax, A., R. H. Griffey, and B. L. Hawkins (1983). *J. Am. Chem. Soc.*, **105**, 7188–7190.

Bell, R. A., J. R. Everett, D. W. Hughes, J. M. Coddington, D. Alkema, P. A. Hader, and T. Neilson (1985). *J. Biomol. Structure and Dynamics*, **2**, 693–707.

Bendall, M. R., D. T. Pegg, and D. M. Doddrell (1983). *J. Magn. Reson.*, **52**, 81–117.

Billeter, M., W. Braun, and K. Wüthrich (1982). *J. Mol. Biol.*, **155**, 321–346.

Billeter, M., M. Engeli, and K. Wüthrich (1985). *Mol. Graphics*, **3**, 79–83; 97–98.

Blumenthal, L. M. (1970). *Theory and Applications of Distance Geometry*, Chelsea, New York.

Bodenhausen, G. and R. R. Ernst (1982). *Mol. Phys.*, **47**, 319–328.

Bodenhausen, G. and D. J. Ruben (1980). *Chem. Phys. Lett.*, **69**, 185–188.

Bodenhausen, G., R. Freeman, and D. L. Turner (1977). *J. Magn. Reson.*, **27**, 511–514.

Bodenhausen, G., R. L. Vold, and R. R. Vold (1980). *J. Magn. Reson.*, **37**, 93–106.

Bodenhausen, G., H. Kogler, and R. R. Ernst (1984a). *J. Magn. Reson.*, **58**, 370–388.

Bodenhausen, G., G. Wagner, M. Rance, O. W. Sørensen, K. Wüthrich, and R. R. Ernst (1984b). *J. Magn. Reson.*, **59**, 542–550.

Boelens, R., P. Gros, R. M. Scheek, J. A. Verpoorte, and R. Kaptein (1985a). *J. Biomol. Structure Dynamics*, **3**, 269–280.

Boelens, R., R. M. Scheek, K. Dijkstra, and R. Kaptein (1985b). *J. Magn. Reson.*, **62**, 378–386.

Bolton, P. H. and G. Bodenhausen (1982). *Chem. Phys. Lett.*, **89**, 139–144.

Borah, B., J. S. Cohen, and A. Bax (1985). *Biopolymers*, **24**, 747–765.

Borer, P. N., L. S. Kan, and P. O. P. Ts'o (1975). *Biochemistry*, **14**, 4847–4863.

Bösch, C., A. Bundi, M. Oppliger, and K. Wüthrich (1978). *Eur. J. Biochem.*, **91**, 209–214.

Bothner-By, A. A. and J. H. Noggle (1979). *J. Am. Chem. Soc.*, **101:18**, 5152–5155.

Bothner-By, A. A., R. L. Stephens, J. Lee, C. D. Warren, and R. W. Jeanholz (1984). *J. Am. Chem. Soc.*, **106**, 811-813.

Bovey, F. A. (1972). *High Resolution NMR of Macrmolecules*, Academic, New York.

Boyd, J., C. M. Dobson, and C. Redfield (1983). *J. Magn. Reson.*, **55**, 170–176.

Boyd, J., C. M. Dobson, and C. Redfield (1985a). *J. Magn. Reson.*, **62**, 543–550.

Boyd, J., C. M. Dobson, and C. Redfield (1985b). *FEBS LETT.*, **186**, 35–40.

Braun, W. and N. Gō (1985). *J. Mol. Biol.*, **186**, 611–626.

Braun, W., C. Bösch, L. R. Brown, N. Gō, and K. Wüthrich (1981). *Biochim. Biophys. Acta*, **667**, 377–396.

Braun, W., G. Wider, K. H. Lee, and K. Wüthrich (1983). *J. Mol. Biol.*, **169**, 921–948.

Braun, W., G. Wagner, E. Wörgötter, M. Vašák, J. H. R. Kägi, and K. Wüthrich (1986). *J. Mol. Biol.*, **187**, 125–129.

Braunschweiler, L. and R. R. Ernst (1983). *J. Magn. Reson.*, **53**, 521–528.

Braunschweiler, L., G. Bodenhausen, and R. R. Ernst (1983). *Mol. Phys.*, **48**, 535–560.

Brennan, R. G. and M. Sundaralingam (1985). *J. Mol. Biol.*, **181**, 561–563.

Broido, M. S. and D. R. Kearns (1982). *J. Am. Chem. Soc.*, **104**, 5207–5216.

Broido, M. S., G. Zon, and T. L. James (1984). *Biochem. Biophys. Res. Comm.*, **119**, 663–670.

Broido, M. S., T. L. James, G. Zon, and J. W. Keepers (1985). *Eur. J. Biochem.*, **150**, 117–128.

Brown, L. R. (1984). *J. Magn. Reson.*, **57**, 513–518.

Brown, L. R. and K. Wüthrich (1981). *Biochim. Biophys. Acta*, **647**, 95–111.

Brown, L. R., C. Bösch, and K. Wüthrich (1981). *Biochim. Biophys. Acta*, **642**, 296–312.

Brown, L. R., W. Braun, Anil-Kumar, and K. Wüthrich (1982). *Biophys. J.*, **37**, 319–328.

Brown, S. C., K. Mullis, C. Levenson, and R. H. Shafer (1984). *Biochemistry*, **23**, 403–408.

Buck, F., H. Rüterjans, R. Kaptein, and K. Beyreuther (1980). *Proc. Natl. Acad. Sci. USA*, **77**, 5145–5148.

Buck, F., K. D. Hahn, W. Zemann, H. Rüterjans, J. R. Sadler, K. Beyreuther, R. Kaptein, R. Scheek, and W. E. Hull (1983). *Eur. J. Biochem.*, **132**, 321–327.

Bundi, A. and K. Wüthrich (1979a). *Biopolymers*, **18**, 285–298.

Bundi, A. and K. Wüthrich (1979b). *Biopolymers*, **18**, 299–312.

Bundi, A., C. Grathwohl, J. Hochmann, R. M. Keller, G. Wagner, and K. Wüthrich (1975). *J. Magn. Reson.*, **18**, 191–198.

Bystrov, V. F. (1976). *Progr. Nucl. Magn. Reson. Spectros.*, **10**, 41–82.

Calladine, C. R. (1982). *J. Mol. Biol.*, **161**, 343–352.

Campbell, I. D., C. M. Dobson, R. J. P. Williams, and A. V. Xavier (1973). *J. Magn. Reson.*, **11**, 172–181.

Chazin, W. J., D. P. Goldenberg, T. E. Creighton, and K. Wüthrich (1985). *Eur. J. Biochem.*, **152**, 429–437.

Chazin, W. J., K. Wüthrich, S. Hyberts, M. Rance, W. A. Denny, and W. Leupin (1986). *J. Mol. Biol.*, in press.

Cheng, D. M. and R. H. Sarma (1977). *J. Am. Chem. Soc.*, **99**, 7333–7348.

Cheng, D. M., L. S. Kan, D. Frechet, P. O. P. Ts'o, S. Uesugi, T. Shida, and M. Ikehara (1984). *Biopolymers*, **23**, 775–795.

Chou, S. H., D. R. Hare, D. E. Wemmer, and B. R. Reid (1983). *Biochemistry*, **22**, 3037–3041.

Chou, S. H., D. E. Wemmer, D. R. Hare, and B. R. Reid (1984). *Biochemistry*, **23**, 2257–2262.

Clore, G. M. and A. M. Gronenborn (1982). *J. Magn. Reson.*, **48**, 402–417.

Clore, G. M. and A. M. Gronenborn (1983). *J. Magn. Reson.*, **53**, 423–442.

Clore, G. M. and A. M. Gronenborn (1984). *Eur. Biophys. J.*, **11**, 95–102.

Clore, G. M. and A. M. Gronenborn (1985a). *EMBO J.*, **4**, 829–835.

Clore, G. M. and A. M. Gronenborn (1985b). *FEBS Lett.*, **179**, 187–198.

Clore, G. M. and A. M. Gronenborn (1985c). *J. Magn. Reson.*, **61**, 158–164.

Clore, G. M., B. J. Kimber, and A. M. Gronenborn (1983). *J. Magn. Reson.*, **54**, 170–173.

Clore, G. M., A. M. Gronenborn, E. A. Piper, L. W. McLaughlin, E. Graesner, and J. H. van Boom (1984). *Biochem. J.*, **221**, 737–751.

Clore, G. M., A. M. Gronenborn, and L. W. McLaughlin (1985a). *Eur. J. Biochem.*, **151**, 153–165.

Clore, G. M., A. M. Gronenborn, D. S. Moss, and I. J. Tickle (1985b). *J. Mol. Biol.*, **185**, 219–226.

Clore, G. M., A. M. Gronenborn, A. T. Brunger, and M. Karplus (1985c). *J. Mol. Biol.* **186**, 435–455.

Connolly, B. A. and F. Eckstein (1984). *Biochemistry*, **23**, 5523–5527.

Crippen, G. M. (1977). *J. Comp. Phys.*, **24**, 96–107.

Cross, T. A. and S. J. Opella (1985). *J. Mol. Biol.*, **182**, 367–381.

Dalgarno, D. C., B. A. Levine, and R. J. P. Williams (1983). *Bioscience Rep.*, **3**, 443–452.

Davis, D. G. and A. Bax (1985). *J. Magn. Reson.*, **64**, 533–535.

Deisenhofer, J. and W. Steigemann (1975). *Acta Cryst.*, **B31**, 238–250.

deLeeuw, H. P. M., C. A. G. Haasnoot, and C. Altona (1980). *Israel J. Chem.*, **20**, 108–126.

Delsuc, M. A., E. Guitter, N. Trotin, and J. Y. Lallemand (1984). *J. Magn. Reson.*, **56**, 163–166.

De Marco, A. and K. Wüthrich (1976). *J. Magn. Reson.*, **24**, 201–204.

De Marco, A., H. Tschesche, G. Wagner, and K. Wüthrich (1977). *Biophys. Struct. Mech.*, **3**, 303–315.

De Marco, A., R. A. Laursen, and M. Llinás (1985). *Biochim. Biophys. Acta*, **827**, 369–380.

Denk, W., G. Wagner, M. Rance, and K. Wüthrich (1985). *J. Magn. Reson.*, **62**, 350–355.

Denk, W., R. Baumann, and G. Wagner (1986). *J. Magn. Reson.*, **67**, 386–390.

Dickerson, R. E. (1983). *Sci. Am.*, **249** (December), 94–111.

Dickerson, R. E., H. R. Drew, B. N. Conner, R. M. Wing, A. V. Fratini, and M. L. Kapka (1982). *Science*, **216**, 475–485.

Dobson, C. M., E. T. Olejniczak, F. M. Poulsen, and R. G. Ratcliffe (1982). *J. Magn. Reson.*, **48**, 97–110.

Dubs, A., G. Wagner, and K. Wüthrich (1979). *Biochim. Biophys. Acta*, **577**, 177–194.

Dwek, R. A. (1973). *Nuclear Magnetic Resonance in Biochemistry: Applications to Enzyme Systems*, Oxford University Press, London.

Dwek, R. A., I. D. Campbell, R. E. Richards, and R. J. P. Williams, Eds. (1977). *NMR in Biology*, Academic, New York.

Ebina, S. and K. Wüthrich (1984). *J. Mol. Biol.*, **179**, 283–288.

Eich, G., G. Bodenhausen, and R. R. Ernst (1982). *J. Am. Chem. Soc.*, **104**, 3731–3732.

Eigen, M. (1964). *Angew. Chem. (Intl. Ed.)*, **3**, 1–19.

Emsley, J. W., J. Feeney, and L. H. Sutcliffe (1966). *High Resolution Nuclear Magnetic Resonance Spectroscopy*, Pergamon, Oxford.

Englander, S. W. and N. R. Kallenbach (1984). *Q. Rev. Biophys.*, **16**, 521–655.

Ernst, R. R. (1966). *Adv. Magn. Reson.*, **2**, 1–135.

Ernst, R. R. (1982). ACS Symposium Series No. 191, 47–61.

Ernst, R. R., G. Bodenhausen, and A. Wokaun (1987). *Principles of Nuclear Magnetic Resonances in One and Two Dimensions*, Oxford University Press, Oxford.

Feeney, J., P. E. Hansen, and G. C. K. Roberts, (1974). *J. Chem. Soc. Chem. Comm.*, 465–466.

Feigon, J., J. M. Wright, W. Leupin, W. A. Denny, and D. R. Kearns (1982). *J. Am. Chem. Soc.*, **104**, 5540–5541.

Feigon, J., W. Leupin, W. A. Denny, and D. R. Kearns (1983). *Biochemistry*, **22**, 5943–5951.

Feigon, J., W. A. Denny, W. Leupin, and D. R. Kearns (1984a). *J. Medicinal Chem.*, **27**, 450–465.

Feigon, J., A. H. J. Wang, G. A. van der Marel, J. H. van Boom, and A. Rich (1984b). *Nucleic Acids Res.*, **12**, 1243–1263.

Figueroa, N., G. Keith, J. L. Leroy, P. Plateau, S. Roy, and M. Gueron (1983). *Proc. Natl. Acad. Sci. USA*, **80**, 4330–4333.

Fischman, A. J., H. R. Wyssbrod, W. C. Agosta, and D. Cowburn (1978). *J. Am. Chem. Soc.*, **100**, 54–58.

Frechet, D., D. M. Cheng, L. S. Kan, and P. O. P. Ts'o (1983). *Biochemistry*, **22**, 5194–5200.

Freeman, R. and G. A. Morris (1979). *Bull. Magn. Reson.*, **1**, 5–26.

Frey, M. H., W. Leupin, O. W. Sørensen, W. A. Denny, R. R. Ernst, and K. Wüthrich (1985a). *Biopolymers*, **24**, 2371–2380.

Frey, M. H., G. Wagner, M. Vašák, O. W. Sørensen, D. Neuhaus, J. H. R. Kägi, R. R. Ernst, and K. Wüthrich (1985b). *J. Am. Chem. Soc.*, **107**, 6847–6851.

Gibbons, W. A., D. Crepaux, J. Delayre, J. J. Dunand, G. Hajdukovic, and H. R. Wyssbrod (1975). "The Study of Peptides by Indor, Differnce NMR and Time-Resolved Double Resonance Techniques," in *Peptides: Chemistry, Structure, Biology*, R. Walter, J. Meienhofer, Eds. Ann Arbor Science, pp. 127–137.

Giessner-Prettre, C. (1984). *J. Biomol. Struct. Dynamics*, **2**, 233–248.

Giessner-Prettre, C. and B. Pullman (1970). *J. Theor. Biol.*, **27**, 87–95.

Glickson, J. D., S. L. Gordon, T. Ph. Pitner, D. G. Agresti, and R. Walter (1976). *Biochemistry*, **15**, 5721–5729.

Gordon, S. L. and K. Wüthrich (1978). *J. Am. Chem. Soc.*, **100**, 7094–7096.

Gorenstein, D. G., K. Lai, and D. O. Shah (1984). *Biochemistry*, **23**, 6717–6723.

Govil, G. and R. V. Hosur (1982). *Conformation of Biological Molecules. New Results from NMR*, Springer, Berlin.

Grathwohl, C. and K. Wüthrich (1974). *J. Magn. Reson.*, **13**, 217–225.

Griesinger, C., O. W. Sørensen, R. R. Ernst (1985). *J. Am. Chem. Soc.*, **107**, 6394–6396.

Griffey, R. H., C. D. Poulter, A. Bax, B. L. Hawkins, Z. Yamaizumi, and S. Nishimura (1983). *Proc. Natl. Acad. Sci. USA*, **80**, 5895–5897.

Griffey, R. H., A. G. Redfield, R. E. Loomis, and F. W. Dahlquist (1985). *Biochemistry*, **24**, 817–822.

Gronenborn, A. M., G. M. Clore, M. Brunori, B. Giardina, G. Falcioni, and M. F. Perutz (1984a). *J. Mol. Biol.*, **178**, 731–742.

Gronenborn, A. M., G. M. Clore, M. B. Jones, and J. Jiricny (1984b). *FEBS Lett.*, **165**, 216–222.

Guéron, M. (1978). *J. Magn. Reson.*, **30**, 515–520.

Guittet, E., M. A. Delsuc, and J. Y. Lallemand (1984). *J. Am. Chem. Soc.*, **106**, 4278–4279.

Gundhi, P., K. V. R. Chary, and R. V. Hosur (1985). *FEBS Lett.*, **191**, 92–96.

Haasnoot, C. A. G. and C. W. Hilbers, (1983). *Biopolymers*, **22**, 1259–1266.

Haasnoot, C. A. G., S. H. de Bruin, R. G. Berendsen, H. G. J. M. Janssen, T. J. J. Binnendijk, C. W. Hilbers, G. A. van der Marel, and J. H. van Boom (1983a). *J. Biol. Struct. Dynamics*, **1**, 115–129.

Haasnoot, C. A. G., H. P. Westerink, G. A. van der Marel, and J. H. van Boom (1983b). *J. Biomol. Struct. Dynamics*, **1**, 131–149.

Haasnoot, C. A. G., H. P. Westerink, G. A. van der Marel, and J. H. van Boom (1984). *J. Biomol. Struct. Dynamics*, **2**, 345–360.

Hahn, E. L. and D. E. Maxwell (1952). *Phys. Rev.*, **88**, 1070–1084.

Harbison, G. S., J. Feigon, D. J. Ruben, J. Herzfeld, and R. G. Griffin (1985). *J. Am. Chem. Soc.*, **107**, 5567–5569.

Hare, D. R. and B. R. Reid (1982a). *Biochemistry,* **21,** 1835–1842.

Hare, D. R. and B. R. Reid (1982b). *Biochemistry,* **21,** 5129–5135.

Hare, D. R., D. E. Wemmer, S. H. Chou, G. Drobny, and B. R. Reid (1983). *J. Mol. Biol.,* **171,** 319–336.

Hare, D. R., N. S. Ribeiro, D. E. Wemmer, and B. R. Reid (1985). *Biochemistry,* **24,** 4300–4306.

Havel, T. F. and K. Wüthrich (1984). *Bull. Math. Biol.,* **46,** 673–698.

Havel, T. F. and K. Wüthrich (1985). *J. Mol. Biol.,* **182,** 281–294.

Havel, T. F., G. M. Crippen, and I. D. Kuntz (1979). *Biopolymers,* **18,** 73–81.

Havel, T. F., I. D. Kuntz, and G. M. Crippen (1983). *Bull. Math. Biol.,* **45,** 665–720.

Heerschap, A., C. A. G. Haasnoot, and C. W. Hilbers (1982). *Nucleic Acids Res.,* **10,** 6981–7000.

Heerschap, A., C. A. G. Haasnoot, and C. W. Hilbers (1983a). *Nucleic Acids Res.,* **11,** 4483–4499.

Heerschap, A., C. A. G. Haasnoot, and C. W. Hilbers (1983b). *Nucleic Acids Res.,* **11,** 4501–4520.

Heerschap, A., J. R. Mellema, H. G. J. M. Janssen, J. A. L. I. Walters, C. A. G. Haasnoot, and C. W. Hilbers (1985). *Eur. J. Biochem.,* **149,** 649–655.

Heus, H. A., J. M. A. van Kimmenade, P. H. van Knippenberg, C. A. G. Haasnoot, S. H. de Bruin, and C. W. Hilbers (1983). *J. Mol. Biol.,* **170,** 939–956.

Hilbers, C. W. (1979). "Hydrogen-Bonded Proton Exchange and its Effect on NMR Spectra of Nucleic Acids," in *Biological Applications of Magnetic Resonance,* R. G. Shulman, Ed., Academic, New York, pp. 1–43.

Hilbers, C. W., A. Heerschap, J. A. L. I. Walters, and C. A. G. Haasnoot (1983a). "NMR Studies of the Structure of Yeast tRNA[Phe] in Solution and of its Complex with the Elongation Factor Tu from B. Stearothermophilus," in *Nucleic Acids: The Vectors of Life,* B. Pullman and J. Jortner, Eds., Reidel, New York, pp. 427–441.

Hilbers, C. W., A. Heerschap, C. A. G. Haasnoot, and J. A. L. I. Walters (1983b). *J. Biomol. Struct. Dynamics,* **1,** 183–207.

Hintermann, M., L. Braunschweiler, G. Bodenhausen, and R. R. Ernst (1982). *J. Magn. Reson.,* **50,** 316–322.

Hoch, J. C., C. M. Dobson, and M. Karplus (1985). *Biochemistry,* **24,** 3831–3841.

Hoffman, R. A. and S. Forsén (1966). *Progr. NMR Spectroscopy,* **1,** 15–204.

Holak, T. A., P. N. Borer, G. C. Levy, J. H. van Boom, and A. H. J. Wang (1984). *Nucleic Acids Res.,* **12,** 4625–4635.

Hore, P. J. (1983). *J. Magn. Reson.,* **55,** 283–300.

Hore, P. J. (1985). *J. Magn. Reson.,* **62,** 561–567.

Hosur, R. V., K. V. R. Chary, and M. Ravikumar (1985a). *Chem. Phys. Lett.,* **116,** 105–108.

Hosur, R. V., M. R. Kumar, K. B. Roy, T. Zu-Kun, H. T. Miles, and G. Govil (1985b). "Strategies of Resonance Assignment and Structure Determination of Oligonucleotides by Two-Dimensional Nuclear Magnetic Resonance," in *Magnetic Resonance in Biology and Medicine,* G. Govil, C. L. Khetrapal, and A. Saran, Eds., Tata McGraw-Hill, New Delhi, pp. 243–260.

Hosur, R. V., M. R. Kumar, and A. Sheth (1985c). *J. Magn. Reson.*, **65**, 375–381.

Hoult, D. I. and R. E. Richards (1975). *Proc. Roy. Soc. London, Ser. A*, **344**, 311–320.

Hull, W. E. and B. D. Sykes (1975). *J. Chem. Phys.*, **63**, 867–880.

Hunkapiller, M. W., M. D. Forgac, E. Ho Yu, and J. H. Richards (1979). *Biochem. Biophys. Res. Comm.*, **87**, 25–31.

Hyde, E. I. and B. R. Reid (1985). *Biochemistry*, **24**, 4307–4314.

Inagaki, F., N. J. Clayden, N. Tamiya, and R. J. P. Williams (1981). *Eur. J. Biochem.*, **120**, 313–322.

IUPAC–IUB Commission on Biochemical Nomenclature (1970). *J. Mol. Biol.*, **52**, 1–17.

IUPAC–IUB Commission on Biochemical Nomenclature (1983). *Eur. J. Biochem.*, **131**, 9–15.

James, T. L. (1975). *Nuclear Magnetic Resonance in Biochemistry*, Academic, New York.

Jamin, N., T. L. James, and G. Zon (1985). *Eur. J. Biochem.*, **152**, 157–166.

Jardetzky, O. and G. C. K. Roberts (1981). *NMR in Molecular Biology*, Academic, New York.

Jeener, J., B. H. Meier, P. Bachmann, and R. R. Ernst (1979). *J. Chem. Phys.*, **71**, 4546–4553.

Johnson, C. E. and F. A. Bovey (1958). *J. Chem. Phys.*, **29**, 1012–1014.

Johnston, P. D. and A. G. Redfield (1977). *Nucleic Acids Res.*, **4**, 3599–3616.

Johnston, P. D. and A. G. Redfield (1978). *Nucleic Acids Res.*, **5**, 3913–3928.

Joseph, A. P. and P. H. Bolton (1984). *J. Am. Chem. Soc.*, **106**, 437–439.

Kabsch, W. and C. Sander (1983). *Biopolymers*, **22**, 2577–2637.

Kainosho, M. and T. Tsuji (1982). *Biochemistry*, **24**, 6273–6279.

Kalbitzer, H. R., R. Leberman, and A. Wittinghofer (1985). *FEBS LETT.*, **180**, 40–42.

Kalk A. and H. J. C. Berendsen (1976). *J. Magn. Reson.*, **24**, 343–366.

Kapka, M. L., C. Yoon, D. Goodsell, P. Pjura, and R. E. Dickerson (1985). *J. Mol. Biol.*, **183**, 553–563.

Kaptein, R., K. Nicolay, and K. Dijkstra (1979). *J. Chem. Soc. Chem. Comm.*, 1092–1094.

Kaptein, R., E. R. P. Zuiderweg, R. M. Scheek, R. Boelens, and W. F. van Gunsteren (1985). *J. Mol. Biol.*, **182**, 179–182.

Karplus, M. (1959). *J. Phys. Chem.*, **30**, 11–15.

Kearns, D. R. (1984). *CRC Critical Reviews in Biochem.*, **15**, 237–290.

Kearns, D. R., D. J. Patel, and R. G. Shulman (1971). *Nature*, **229**, 338–339.

Keepers, J. W. and T. L. James (1984). *J. Magn. Reson.*, **57**, 404–426.

Keim, P., R. A. Vigna, R. C. Marshall, and F. R. N. Gurd (1973a). *J. Biol. Chem.*, **248**, 6104–6113.

Keim, P., R. A. Vigna, J. S. Morrow, R. C. Marshall, and F. R. N. Gurd (1973b). *J. Biol. Chem.*, **248**, 7811-7818.

Keller, R. M. and K. Wüthrich (1978). *Biochim. Biophys. Acta*, **533**, 195–208.

Keller, R. M. and K. Wüthrich (1980). *Biochim. Biophys. Acta*, **621**, 204–217.

Keller, R. M. and K. Wüthrich (1981). "Multiple Irradiation ^{1}H NMR Experiments with Hemoproteins," in *Biological Magnetic Resonance*, L. J. Berliner and J. Reuben, Eds., Vol. 3, Plenum, New York and London, pp. 1–52.

Keller, R. M., R. Baumann, E. H. Hunziker-Kwik, F. J. Joubert, and K. Wüthrich (1983). *J. Mol. Biol.*, **163**, 623–646.

Kessler, H., M. Bernd, H. Kogler, J. Zarbock, O. W. Sørensen, G. Bodenhausen, and R. R. Ernst (1983). *J. Am. Chem. Soc.*, **105**, 6944-6952.

Kime, M. J. and P. B. Moore (1983). *Biochemistry*, **22**, 2615–2622.

King, G. and P. E. Wright (1982). *Biochem. Biophys. Res. Comm.*, **106**, 559–565.

King, G. and P. E. Wright (1983). *J. Magn. Reson.*, **54**, 328–332.

Klevit, R. E. (1985). *J. Magn. Reson.*, **62**, 551–555.

Kline, A. D. and K. Wüthrich (1985). *J. Mol. Biol.*, **183**, 503–507.

Kline, A. D. and K. Wüthrich (1986). *J. Mol. Biol.*, submitted.

Kline, A. D., W. Braun, and K. Wüthrich (1986). *J. Mol. Biol.*, **189**, 377–382.

Kojiro, C. L. and J. L. Markley (1983). *FEBS Lett.*, **162**, 52–56.

Kuntz, I. D., G. M. Crippen, P. A. Kollman, and D. Kimmelman (1976). *J. Mol. Biol.*, **106**, 983–994.

Kuntz, I. D., G. M. Crippen, and P. A. Kollman (1979). *Biopolymers*, **18**, 939–957.

Kuo, M. C. and W. A. Gibbons (1980). *Biophys. J.*, **32**, 807–836.

Lai, K., D. O. Shah, E. De Rose, and D. G. Gorenstein (1984). *Biochem. Biophys. Res. Comm.*, **121**, 1021–1026.

La Mar, G. N. (1979). "Model Compounds as Aids in Interpreting NMR Spectra of Hemoproteins," in *Biological Applications of Magnetic Resonance*, R. G. Shulman, Ed., Academic, New York, pp. 305–343.

Lane, A. N. and O. Jardetzky (1985). *Eur. J. Biochem.*, **152**, 395–404; 405–409; 411–418.

Laue, E. D., J. Skilling, J. Staunton, S. Sibisi, and R. G. Brereton (1985). *J. Magn. Reson.*, **62**, 437–452.

Lankhorst, P. P., C. A. G. Haasnoot, C. Erkelens, and C. Altona (1984). *Nucleic Acids Res.*, **12**, 5419–5428.

Lankhorst, P. P., G. A. van der Marel, G. Wille, J. H. van Boom, and C. Altona (1985). *Nucleic Acids Res.*, **13**, 3317-3333.

Leach, S. J., G. Némethy, and H. A. Scheraga (1977). *Biochem. Biophys. Res. Comm.*, **75**, 207–215.

Lee, S. J., H. Akutsu, Y. Kyogoku, K. Kitano, Z. Tozuka, A. Ohta, E. Ohtsuka, and M. Ikehara (1985). *J. Biochem.*, **98**, 1463–1472.

Lefevre, J. F., A. N. Lane, and O. Jardetzky (1985). *J. Mol. Biol.*, **185**, 689–699.

LeMaster, D. M. and F. M. Richards (1982). *Anal. Biochem.*, **122**, 238–247.

Live, D. H., D. G. Davis, W. C. Agosta, and D. Cowburn (1984). *J. Am. Chem. Soc.*, **106**, 6104–6105.

Live, D., I. M. Armitage, D. C. Dalgarno, and D. Cowburn (1985). *J. Am. Chem. Soc.*, **107**, 1775–1777.

Macura, S. and R. R. Ernst (1980). *Mol. Phys.*, **41**, 95–117.

Macura, S., Y. Huang, D. Suter, and R. R. Ernst (1981). *J. Magn. Reson.*, **43**, 259–281.

Macura, S., K. Wüthrich, and R. R. Ernst (1982a). *J. Magn. Reson.*, **46**, 269–282.

Macura, S., K. Wüthrich, and R. R. Ernst (1982b). *J. Magn. Reson.*, **47**, 351–357.

Macura, S., N. G. Kumar, and L. R. Brown (1983). *Biochem. Biophys. Res. Comm.*, **117**, 486–492.

Marion, D. (1985). *FEBS Lett.*, **192**, 99–103.

Marion, D. and G. Lancelot (1984). *Biochem. Biophys. Res. Comm.*, **124**, 774–783.

Marion, D. and K. Wüthrich (1983). *Biochem. Biophys. Res. Comm.*, **113**, 967–974.

Markley, J. L. (1979). "Catalytic Groups of Serine Proteinases. NMR Investigations," in *Biological Applications of Magnetic Resonance*, R. G. Shulman, Ed., Academic, New York, pp. 397–461.

Markley, J. L. and E. L. Ulrich (1984). *Ann. Rev. Biophys. Bioeng.*, **13**, 493–521.

Markley, J. L., I. Putter, and O. Jardetzky (1968). *Science*, **161**, 1249–1251.

McDonald, C. C. and W. D. Phillips (1967). *J. Am. Chem. Soc.*, **89**, 6332–6341.

Mehlkopf, A. F., D. Korbee, T. A. Tiggelman, and R. Freeman (1984). *J. Magn. Reson.*, **58**, 315–323.

Meier, B. H. and R. R. Ernst (1979). *J. Am. Chem. Soc.*, **101**, 6441–6442.

Meier, B. U., G. Bodenhausen, and R. R. Ernst (1984). *J. Magn. Reson.*, **60**, 161–163.

Mellema, J. R., A. K. Jellema, C. A. G. Haasnoot, J. H. van Boom, and C. Altona (1984). *Eur. J. Biochem.*, **141**, 165–175.

Momany, F. A., R. F. McGuire, A. W. Burgess, and H. A. Scheraga (1975). *J. Phys. Chem.*, **79**, 2361–2381.

Moore, G. R., M. N. Robinson, G. Williams, and R. J. P. Williams (1985). *J. Mol. Biol.*, **183**, 429–446.

Moore, P. B., J. A. Langer, B. P. Schoenborn, and D.M. Engelman (1977). *J. Mol. Biol.*, **112**, 199–234.

Moore, P. B. and E. Weinstein (1979). *J. Appl. Cryst.*, **12**, 321–326.

Müller, L. (1979). *J. Am. Chem. Soc.*, **101**, 4481–4484.

Müller, N., G. Bodenhausen, K. Wüthrich, and R. R. Ernst (1985). *J. Magn. Reson.*, **65**, 531–534.

Müller, N., R. R. Ernst, und K. Wüthrich (1986). *J. Am. Chem. Soc.*, in press.

Nagayama, K. (1981). *Adv. Biophys.*, **14**, 139–204.

Nagayama, K. and K. Wüthrich (1981a). *Eur. J. Biochem.*, **114**, 365–374.

Nagayama, K. and K. Wüthrich (1981b). *Eur. J. Biochem.*, **115**, 653–657.

Nagayama, K., K. Wüthrich, P. Bachmann, and R. R. Ernst (1977). *Biochem. Biophys. Res. Comm.*, **78**, 99–105.

Nagayama, K., P. Bachmann, K. Wüthrich, and R. R. Ernst (1978). *J. Magn. Reson.*, **31**, 133–148.

Nagayama, K., K. Wüthrich, and R. R. Ernst (1979). *Biochem. Biophys. Res. Comm.*, **90**, 305–311.

Nagayama, K., Anil-Kumar, K. Wüthrich, and R. R. Ernst (1980). *J. Magn. Reson.*, **40**, 321–334.

Nagayama, K., Y. Kobayashi, and Y. Kyogoku (1983). *J. Magn. Reson.*, **51**, 84–94.

Neidig, K. P., H. Bodenmueller, and H. R. Kalbitzer (1984). *Biochem. Biophys. Res. Comm.*, **125**, 1143–1150.

Némethy, G., M. S. Pottle, and H. A. Scheraga (1983). *J. Phys. Chem.*, **87**, 1883–1887.

Neuhaus, D., G. Wagner, M. Vašák, J. H. R. Kägi, and K. Wüthrich (1984a). *Eur. J. Biochem.*, **143**, 659–667.

Neuhaus, D., G. Wider, G. Wagner, and K. Wüthrich (1984b). *J. Magn. Reson.*, **57**, 164–168.

Neuhaus, D., G. Wagner, M. Vašák, J. H. R. Kägi, and K. Wüthrich (1985). *Eur. J. Biochem.*, **151**, 257–273.

Noggle, J. H. and R. E. Schirmer (1971). *The Nuclear Overhauser Effect*, Academic, New York.

Olejniczak, E. T., C. M. Dobson, M. Karplus, and R. M. Levy (1984a). *J. Am. Chem. Soc.*, **106**, 1923–1930.

Olejniczak, E. T., F. M. Poulsen, and C. M. Dobson (1984b). *J. Magn. Reson.*, **59**, 518–523.

Opella, S. J. and P. Lu, Eds. (1979). *NMR and Biochemistry—A Symposium Honoring Mildred Cohn*, Dekker, New York.

Oschkinat, H. and R. Freeman (1984). *J. Magn. Reson.*, **60**, 164–169.

Otting, G. and K. Wüthrich (1986). *J. Magn. Reson.*, **66**, 359–363.

Otting, G., H. Widmer, G. Wagner, and K. Wüthrich (1986). *J. Magn. Reson.*, **66**, 187–193.

Otvos, J. D., H. R. Engeseth, and S. Wehrli (1985). *J. Magn. Reson.*, **61**, 579–584.

Pardi, A., R. Walker, H. Rapoport, G. Wider, and K. Wüthrich (1983a). *J. Am. Chem. Soc.*, **105**, 1652–1653.

Pardi, A., G. Wagner, and K. Wüthrich (1983b). *Eur. J. Biochem.*, **137**, 445–454.

Pardi, A., K. M. Morden, D. J. Patel, and I. Tinoco, Jr. (1983c). *Biochemistry*, **22**, 1107–1113.

Pardi, A., M. Billeter, and K. Wüthrich (1984). *J. Mol. Biol.*, **180**, 741–751.

Patel, D. J. (1979). *Acc. Chem. Res.*, **12**, 118–125.

Patel, D. J. (1982). *Proc. Natl. Acad. Sci. USA*, **79**, 6424–6428.

Patel, D. J., A. Pardi, and K. Itakura (1982a). *Science*, **216**, 581–590.

Patel, D. J., S. A. Kozlowski, A. Nordheim, and A. Rich (1982b). *Proc. Natl. Acad. Sci. USA*, **79**, 1413–1417.

Patel, D. J., S. Ikuta, S. A. Kozlowski, and K. Itakura (1983). *Proc. Natl. Acad. Sci. USA*, **80**, 2184–2188.

Patel, D. J., S. A. Kozlowsky, D. R. Hare, B. Reid, S. Ikuta, N. Lander, and K. Itakura (1985a). *Biochemistry*, **24**, 926–935.

Patel, D. J., S. A. Kozlowski, M. Weiss, and R. Bhatt (1985b). *Biochemistry*, **24**, 936–944.

Perkins, S. J. (1982). "Application of Ring Current Calculations to the Proton NMR of Proteins and Transfer RNA," *Biological Magnetic Resonance*, L. J. Berliner and J. Reuben, Eds., Vol. 4, Plenum Press, New York, pp. 79–144.

Perkins, S. J. and K. Wüthrich (1979). *Biochim. Biophys. Acta*, **576**, 409–423.

Petersheim, M., S. Mehdi, and J. A. Gerlt (1984). *J. Am. Chem. Soc.*, **106**, 439–440.

Pfändler, P., G. Bodenhausen, B. U. Meier, and R. R. Ernst (1985). *Anal. Chem.*, **57**, 2510–2516.

Piantini, U., O. W. Sørensen, and R. R. Ernst (1982). *J. Am. Chem. Soc.*, **104**, 6800–6801.

Plateau, P. and M. Guéron (1982). *J. Am. Chem. Soc.*, **104**, 7310–7311.

Pohl, F. M. and T. M. Jovin (1972). *J. Mol. Biol.*, **67**, 375–396.

Pople, J. A., W. G. Schneider, and H. J. Bernstein (1959). *High Resolution NMR*, McGraw-Hill, New York.

Pouzard, G., S. Sukumar and L. D. Hall (1981). *J. Am. Chem. Soc.*, **103**, 4209–4215.

Prestegard, J. H. and J. N. Scarsdale (1985). *J. Magn. Reson.*, **62**, 136–140.

Ramachandran, G. N. and V. Sasisekharan (1968). *Adv. Protein Chem.*, **23**, 283–437.

Rance, M., O. W. Sørensen, G. Bodenhausen, G. Wagner, R. R. Ernst, and K. Wüthrich (1983). *Biochem. Biophys. Res. Comm.*, **117**, 479–485.

Rance, M., G. Wagner, O. W. Sørensen, K. Wüthrich, and R. R. Ernst (1984). *J. Magn. Reson.*, **59**, 250–261.

Rance, M., O. W. Sørensen, W. Leupin, H. Kogler, K. Wüthrich, and R. R. Ernst (1985a). *J. Magn. Reson.*, **61**, 67–80.

Rance, M., G. Bodenhausen, G. Wagner, K. Wüthrich, and R. R. Ernst (1985b). *J. Magn. Reson.*, **62**, 497–510.

Rance, M., C. Dalvit, and P. E. Wright (1985c). *Biochem. Biophys. Res. Comm.*, **131**, 1094–1102.

Redfield, A. G. and R. K. Gupta (1971). *Cold Spring Harbor Symp. Quant. Biol.*, **36**, 405–411.

Redfield, A. G. and S. D. Kunz (1975). *J. Magn. Reson.*, **19**, 250–254.

Redfield, A. G. and S. D. Kunz (1979). "Proton Resonance Spectrometer for Biochemical Applications," in *NMR and Biochemistry*, S. J. Opella and P. Lu, Eds., Dekker, New York, pp. 225-239.

Redfield, C., J. C. Hoch, and C. M. Dobson (1983). *FEBS Lett.*, **159**, 132–136.

Reid, D. G., S. A. Salisbury, S. Bellard, Z. Shakked, and D. H. Williams (1983). *Biochemistry*, **22**, 2019–2025.

Rich, A. and U. L. RajBhandary (1976). *Ann. Rev. Biochem.*, **45**, 805–860.

Richardson, J. (1981). *Adv. Protein Chem.*, **34**, 167–339.

Richarz, R. and K. Wüthrich (1978a). *Biopolymers*, **17**, 2133-2141.

Richarz, R. and K. Wüthrich (1978b). *J. Magn. Reson.*, **30**, 147-150.

Richarz, R., H. Tschesche, and K. Wüthrich (1980). *Biochemistry*, **19**, 5711-5715.

Robillard, G. T. and B. R. Reid (1979). "Elucidation of Nucleic Acid Structure by Proton NMR," in *Biological Applications of Magnetic Resonance*, R. G. Shulman, Ed., Academic, New York, pp. 45–112.

Roder, H., G. Wagner, and K. Wüthrich (1985). *Biochemistry*, **24**, 7396–7407; 7407–7411.

Roth, K., B. J. Kimber, and J. Feeney (1980). *J. Magn. Reson.*, **41**, 302–309.

Roy, S., M. Z. Papastavros, and A. G. Redfield (1982a). *Biochemistry*, **21**, 6081–6088.

Roy, S., M. Z. Papastavros, and A. G. Redfield (1982b). *Nucleic Acids Res.*, **10**, 8341–8349.

Saenger, W. (1984). *Principles of Nucleic Acid Structure*, Springer, Berlin.

Sanderson, M. R., J. R. Mellema, G. A. van der Marel, G. Wille, J. H. van Boom, and C. Altona (1983). *Nucleic Acids Res.*, **11**, 3333–3346.

Scheek, R. M., N. Russo, R. Boelens, R. Kaptein, and J. H. van Boom (1983). *J. Am. Chem. Soc.*, **105**, 2914–2916.

Scheek, R. M., R. Boelens, N. Russo, J. H. van Boom, and R. Kaptein (1984). *Biochemistry*, **23**, 1371–1376.

Schejter, E., S. Roy, V. Sánchez, and A. G. Redfield (1982). *Nucleic Acids Res.*, **10**, 8297–8304.

Schulz, G. E. and R. H. Schirmer (1979). *Principles of Protein Structure*, Springer, Berlin.

Schwartz, A. L. and J. D. Cutnell (1983). *J. Magn. Reson.*, **53**, 398–411.

Senn, H., M. Billeter, and K. Wüthrich (1984). *Eur. Biophys. J.*, **11**, 3–5.

Shaka, A. J. and R. Freeman (1983). *J. Magn. Reson.*, **51**, 169–173.

Shulman, R. G., Ed. (1979). *Biological Applications of Magnetic Resonance*, Academic, New York.

Sibanda, B. L., and J. M. Thornton (1985). *Nature*, **316**, 170–174.

Sippl, M. J. and H. A. Scheraga (1985). *Proc. Natl. Acad. Sci. USA*, **82**, 2197–2201.

Solomon, I. (1955). *Phys. Rev.*, **99**, 559–565.

Sørensen, O. W. (1984). Modern Pulse Techniques in Liquid State Nuclear Magnetic Resonance Spectroscopy, Ph.D. Thesis, Eidgenössische Technische Hochschule Zürich, ADAG Druck AG, Zürich.

Sørensen, O. W., G. W. Eich, M. H. Levitt, G. Bodenhausen, and R. R. Ernst (1983). *Progr. Nucl. Magn. Reson. Spectros.*, **16**, 163–192.

Sørensen, O. W., M. Rance, and R. R. Ernst (1984). *J. Magn. Reson.*, **56**, 527–534.

Sørensen, O. W., C. Griesinger, and R. R. Ernst (1985). *J. Am. Chem. Soc.*, **107**, 7778–7779.

Stassinopoulou, C. I., G. Wagner, and K. Wüthrich (1984). *Eur. J. Biochem.*, **145**, 423–430.

States, D. J., R. A. Haberkorn, and D. J. Ruben (1982). *J. Magn. Reson.*, **48**, 286–292.

States, D. J., C. M. Dobson, M. Karplus, and T. E. Creighton (1984). *J. Mol. Biol.*, **174**, 411–418.

Stothers, J. B. (1972). *Carbon-13 NMR Spectroscopy*, Academic, New York.

Štrop, P. and K. Wüthrich (1983). *J. Mol. Biol.*, **166**, 631–640.

Štrop, P., G. Wider, and K. Wüthrich (1983a). *J. Mol. Biol.*, **166**, 641–667.

Štrop, P., D. Čechová, and K. Wüthrich (1983b). *J. Mol. Biol.*, **166**, 669–676.

Tran-Dinh, S., J. M. Neumann, J. Taboury, T. Huynh-Dinh, S. Renous, B. Genissel, and J. Igolen (1983). *Eur. J. Biochem.*, **133**, 579–589.

Tropp, J. and A. G. Redfield (1981). *Biochemistry*, **20**, 2133–2140.

Tüchsen, E. and C. Woodward (1985a). *J. Mol. Biol.*, **185**, 405–419.

Tüchsen, E. and C. Woodward (1985b). *J. Mol. Biol.*, **185**, 421–430.

Ulrich, E. L., E. M. M. John, G. R. Gough, M. J. Brunden, P. T. Gilham, W. M. Westler, and J. L. Markley (1983). *Biochemistry*, **22**, 4362–4365.

van de Ven, F. J. M., S. H. de Bruin, and C. W. Hilbers (1984). *FEBS Lett.*, **169**, 107–111.

van de Ven, F. J. M., C. A. G. Haasnoot, and C. W. Hilbers (1985). *J. Magn. Reson.*, **61**, 181–187.

Wagner, G. (1983a). *J. Magn. Reson.*, **55**, 151–156.

Wagner, G. (1983b). *Quart. Rev. Biophys.*, **16**, 1–57.

Wagner, G. (1984). *J. Magn. Reson.*, **57**, 497–505.

Wagner, G. and K. Wüthrich (1979). *J. Magn. Reson.*, **33**, 675–680.

Wagner, G. and K. Wüthrich (1982a). *J. Mol. Biol.*, **155**, 347–366.

Wagner, G. and K. Wüthrich (1982b). *J. Mol. Biol.*, **160**, 343–361.

Wagner, G. and E. R. P. Zuiderwg (1983). *Biochem. Biophys. Res. Comm.*, **113**, 854–860.

Wagner, G., K. Wüthrich, and H. Tschesche (1978). *Eur. J. Biochem.*, **86**, 67–76.

Wagner, G., Anil-Kumar, and K. Wüthrich (1981). *Eur. J. Biochem.*, **114**, 375–384.

Wagner, G., A. Pardi, and K. Wüthrich (1983). *J. Am. Chem. Soc.*, **105**, 5948–5949.

Wagner, G., C. I. Stassinopoulou, and K. Wüthrich (1984). *Eur. J. Biochem.*, **145**, 431–436.

Wagner, G., G. Bodenhausen, N. Müller, M. Rance, O. W. Sørensen, R. R. Ernst, and K. Wüthrich (1985a). *J. Am. Chem. Soc.*, **107**, 6440–6446.

Wagner, G., D. Neuhaus, E. Wörgötter, M. Vašák, J. R. H. Kägi, and K. Wüthrich (1986a). *J. Mol. Biol.*, **187**, 131–135.

Wagner, G., D. Neuhaus, M. Vašák, E. Wörgötter, J. H. R. Kägi, and K. Wüthrich (1986b). *Eur. J. Biochem.*, in press.

Wako, H. and H. A. Scheraga (1981). *Macromolecules*, **14**, 961–969.

Wang, A. H. J., G. J. Quigley, F. J. Kolpak, G. S. van der Marel, J. H. S. van Boom, and A. Rich (1981). *Science*, **211**, 171–176.

Wang, A. H. J., T. Hakoshima, G. van der Marel, J. H. van Boom, and A. Rich (1984). *Cell*, **37**, 321–331.

Wang, A. H. J., R. V. Gessner, G. A. van der Marel, J. H. van Boom, and A. Rich (1985). *Proc. Natl. Acad. Sci. USA*, **82**, 3611-3615.

Weber, P. L., G. Drobny, and B. R. Reid (1985a). *Biochemistry*, **24**, 4549–4552.

Weber, P. L., D. E. Wemmer, and B. R. Reid (1985b). *Biochemistry*, **24**, 4553–4562.

Weiss, M. A., D. J. Patel, R. T. Sauer, and M. Karplus (1984a). *Proc. Natl. Acad. Sci. USA*, **81**, 130–134.

Weiss, M. A., D. J. Patel, R. T. Sauer, and M. Karplus (1984b). *Nucleic Acids Res.*, **12**, 4035–4047.

Wemmer, D. E. and N. R. Kallenbach (1983). *Biochemistry*, **22**, 1901–1906.

Wemmer, D. E., S. H. Chou, and B. R. Reid (1984). *J. Mol. Biol.*, **180**, 41–60.

Wemmer, D. E., K. S. Srivenugopal, B. R. Reid, and D. R. Morris (1985). *J. Mol. Biol.*, **185**, 457–459.

Westerink, H. P., G. A. van der Marel, J. H. van Boom, and C. A. G. Haasnoot (1984). *Nucleic Acids Res.*, **12**, 4323–4338.

Wider, G., R. Baumann, K. Nagayama, R. R. Ernst, and K. Wüthrich (1981). *J. Magn. Reson.*, **42**, 73–87.

Wider, G., K. H. Lee, and K. Wüthrich (1982). *J. Mol. Biol.*, **155**, 367–388.

Wider, G., R. V. Hosur, and K. Wüthrich (1983). *J. Magn. Reson.*, **52**, 130–135.

Wider, G., S. Macura, Anil-Kumar, R. R. Ernst, and K. Wüthrich (1984). *J. Magn. Reson.*, **56**, 207–234.

Williamson, M. P., D. Marion, and K. Wüthrich (1984). *J. Mol. Biol.*, **173**, 341–359.

Williamson, M. P., T. F. Havel, and K. Wüthrich (1985). *J. Mol. Biol.*, **182**, 295–315.

Woodward, C. K., I. Simon, and E. Tüchsen (1982). *Mol. Cell. Biochem.*, **48**, 135–160.

Wüthrich, K. (1970). *Struct. Bonding*, (Berlin), **8**, 53–121.

Wüthrich, K. (1976). *NMR in Biological Research: Peptides and Proteins*, North Holland, Amsterdam.

Wüthrich, K. (1983). *Biopolymers*, **22**, 131–138.

Wüthrich, K. and G. Wagner (1978). *Trends Biochem. Sci.*, **3**, 227–230.

Wüthrich, K. and G. Wagner (1979). *J. Mol. Biol.*, **130**, 1–18.

Wüthrich, K. and G. Wagner (1984). *Trends Biochem. Sci.*, **9**, 152–154.

Wüthrich, K., G. Wagner, R. Richarz, and S. J. Perkins (1978). *Biochemistry*, **17**, 2253–2263.

Wüthrich, K., K. Nagayama, and R. R. Ernst (1979). *Trends Biochem. Sci.*, **4**, N178–N181.

Wüthrich, K., G. Wider, G. Wagner, and W. Braun (1982). *J. Mol. Biol.*, **155**, 311–319.

Wüthrich, K., M. Billeter, and W. Braun (1983). *J. Mol. Biol.*, **169**, 949–961.

Wüthrich, K., M. Billeter, and W. Braun (1984a). *J. Mol. Biol.*, **180**, 715–740.

Wüthrich, K., P. Štrop, S. Ebina, and M. P. Williamson (1984b). *Biochem. Biophys. Res. Comm.*, **122**, 1174–1178.

Zuiderweg, E. R. P., R. Kaptein, and K. Wüthrich (1983a). *Proc. Natl. Acad. Sci. USA*, **80**, 5837–5841.

Zuiderweg, E. R. P., R. Kaptein, and K. Wüthrich (1983b). *Eur. J. Biochem.*, **137**, 279–293.

Zuiderweg, E. R. P., M. Billeter, R. Boelens, R. M. Scheek, K. Wüthrich, and R. Kaptein (1984a). *FEBS Lett.*, **174**, 243–247.

Zuiderweg, E. R. P., M. Billeter, R. Kaptein, R. Boelens, R. M. Scheek, and K. Wüthrich (1984b). "Solution Conformation of *E. Coli Lac* Repressor DNA Binding domain by 2D NMR: Sequence Location and Spatial Arrangement of three α-helices," in *Progress in Bioorganic Chemistry and Molecular Biology*, Yu. A. Ovchinnikov Ed., Elsevier, Amsterdam, pp. 65–70.

Zuiderweg, E. R. P., R. Boelens, and R. Kaptein (1985). *Biopolymers*, **24**, 601–611.

Index